FUNGI: BIOLOGICAL DIVERSITY

FUNGI: BIOLOGICAL DIVERSITY

C.S. Chandoliya

CYBER TECH PUBLICATIONS
4264/3, Ansari Road, Daryaganj, New Delhi-110002 (India)
Ph.: 011-23244078. Fax: 011-23280028
E-Mail: cyberpublicationsdelhi@yahoo.com

FUNGI: BIOLOGICAL DIVERSITY

C.S. Chandoliya

First Edition 2009

Published by :

G.S. Rawat for **Cyber Tech Publications**
4264/3, Ansari Road, Daryaganj, New Delhi-110002 (India)
Ph.: 011-23244078. Fax: 011-23280028
E-Mail: cyberpublicationsdelhi@yahoo.com

Printed at : **Amit Enterprises**, Maujpur, Delhi, (9810878184)

Preface

The balance between people and their environment is being upset. Escalating human numbers and increasing demands for material resources are leading to the transformation and degradation of ecosystems worldwide, with consequent loss of genetic diversity and an inevitable rise in the extinction of species. More and more land is being converted to intensive production of food, timber and other plant products, much of the world's pasture land is overgrazed, and soil erosion and salinization are reducing the fertility of farmlands, especially in semi-arid regions. Wild species are being displaced or overcropped, and wild ecosystems can no longer be taken for granted as reservoirs of genetic diversity and regulators of the cycles of the elements.

Plants are a central component of this threatened nature. The loss of plants is very significant, for they stand at the base of food webs and provide habitats for other organisms. Although people in some modern societies may be mentally distanced from the reality of nature and are unaware of the services it provides, even city inhabitants are part of wider ecosystems, based on wild plants and natural vegetation. For example, genes from the wild can play major roles in the breeding of new varieties of food crops and other cultivated plants. New medicines continue to be derived from wild species. In many countries, natural and semi-natural ecosystems provide many plant products essential for human welfare, including fuelwood, timber, fiber, medicinal plants, fruits and nuts.

— *Author*

CONTENTS

Introduction to the Fungi

The Kingdom Fungi includes some of the most important organisms, both in terms of their ecological and economic roles. By breaking down dead organic material, they continue the cycle of nutrients through ecosystems. In addition, most vascular plants could not grow without the symbiotic fungi, or mycorrhizae, that inhabit their roots and supply essential nutrients. Other fungi provide numerous drugs (such as penicillin and other antibiotics), foods like mushrooms, truffles and morels, and the bubbles in bread, champagne, and beer.

Fungi also cause a number of plant and animal diseases: in humans, ringworm, athlete's foot, and several more serious diseases are caused by fungi. Because fungi are more chemically and genetically similar to animals than other organisms, this makes fungal diseases very difficult to treat. Plant diseases caused by fungi include rusts, smuts, and leaf, root, and stem rots, and may cause severe damage to crops. However, a number of fungi, in particular the yeasts, are important "model organisms" for studying problems in genetics and molecular biology.

FOSSIL RECORD

While fungi are not uncommon fossils, their fossils have not received a great deal of attention compared to other groups of fossils. Their fossils tend to be microscopic; very few large fungal bodies, such as mushrooms, have ever been found as fossils. Fossil fungi are often difficult or impossible to identify. The fungal filaments shown above at left are a case in point; found in Cretaceous amber from north France, they resemble living filaments of the common ascomycete *Candida*; however, since there is little information on how this fossil organism lived or how it reproduced (both important in recognizing modern taxa), its true affinities may never be known. By contrast, the Miocene fossil at right above has preserved the **perithecium**, an enclosed reproductive structure. Features of the spores and the perithecium in which they occur suggest that this may be a fossil species of *Savoryella*.

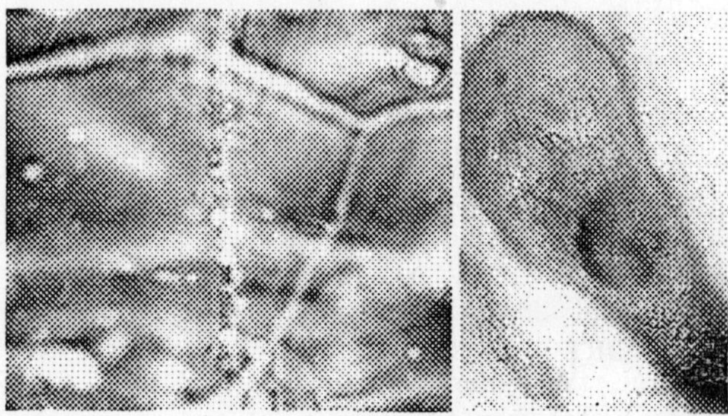

Fossil Fungi : At left are fossil hyphae from the Cretaceous of northern France. The filaments resemble those of the living genus *Candida*. At right is a Miocene perithecium from Nevada. The fine preservation is due to the silicification of chert in which it was embedded.

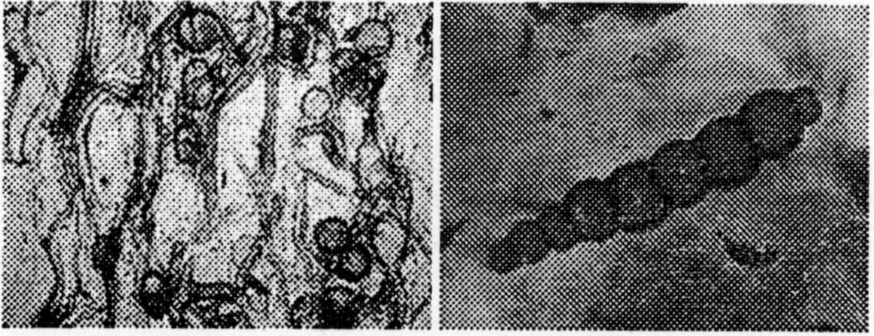

More Fossil Fungi : At left is a section through a silicified stem of *Aglaophyton* from the Devonian Rhynie Chert. Among the cells are a fossilized fungus *Paleomyces*. At right are fungal spores from Lost Chicken Creek.

Recent careful studies of some well-preserved material have contributed much to our knowledge of fossil fungi. In particular, microscopic examination of fossil fungi from the Devonian-age

Rhynie Chert in Aberdeenshire, Scotland, has shown that fungi and land plants were forming symbiotic relationships even at that very early stage in terrestrial evolution. In fact, all four major groups of modern fungi have now been found in Devonian strata, showing that the fungi had successfully invaded the land and begun to diversify before the first vertebrates crawled out of the sea!

Tho oldest fossil fungi so far known are probably chytrid-like forms from the Vendian Period (Late Precambrian), found in north Russia. Older fossils of Precambrian "fungi" are now usually considered to be empty sheaths of filamentous cyanobacteria, or else are not distinct enough to be placed in any taxon with certainty. Fossil fungi older than the Devonian are rare; the fungi may have undergone an evolutionary radiation at about the same time that the land plants began to radiate.

LIFE HISTORY AND ECOLOGY

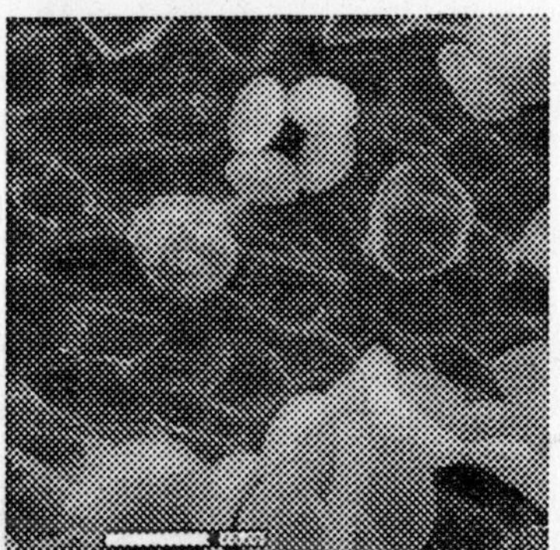

This photograph taken using the UCMP Environmental Scanning Electron Microscope

Fungi exist primarily as filamentous dikaryotic organisms.

As part of their life cycle, fungi produce spores. In this electron micrograph of a mushroom gill, the four spores produced by meiosis (seen in the center of this picture) are carried on a clublike **sporangium** (visible to the left and right). From these spores, haploid **hyphae** grow and ramify, and may give rise to asexual sporangia, special hyphae which produce spores without meiosis.

The sexual phase is begun when haploid hyphae from two different fungal organisms meet and fuse. When this occurs, the cytoplasm from the two cells fuses, but the nuclei remain separate and distinct. The single hypha produced by fusion typically has two nuclei per "cell", and is known as a **dikaryon**, meaning "two nuclei". The dikaryon may live and grow for years, and some are thought to be many centuries old. Eventually, the dikaryon forms sexual sporangia in which the nuclei fuse into one, which then undergoes meiosis to form haploid spores, and the cycle is repeated.

Some fungi, especially the chytrids and zygomycetes, have a life cycle more like that found in many protists. The organism is haploid, and has no diploid phase, except for the sexual sporangium. A number of fungi have lost the capacity for sexual reproduction, and reproduce by asexual spores or by vegetative growth only. These fungi are referred to as Fungi Imperfecti, and include, among other members, the athlete's foot and the fungus in bleu cheese. Other fungi, such as the yeasts, primarily reproduce through asexual **fission**, or by **fragmentation** — breaking apart, with each of the pieces growing into a new organism.

Fungi are Heterotrophic

Fungi are not able to ingest their food like animals do, nor can they manufacture their own food the way plants do. Instead, fungi feed by **absorption** of nutrients from the environment around them. They accomplish this by growing through and within the **substrate** on which they are feeding. Numerous hyphae network through the wood, cheese, soil, or flesh from which they are growing. The hyphae secrete digestive enzymes which break down the substrate, making it easier for the fungus to absorb the nutrients which the substrate contains.

This filamentous growth means that the fungus is in intimate contact with its surroundings; it has a very large surface area compared to its volume. While this makes diffusion of nutrients into the hyphae easier, it also makes the fungus susceptible to dessication and ion imbalance. But usually this is not a problem, since the fungus is growing within a moist substrate.

Most fungi are **saprophytes**, feeding on dead or decaying material. This helps to remove leaf litter and other debris that would otherwise accumulate on the ground. Nutrients absorbed by the fungus then become available for other organisms which may eat fungi. A very few fungi actively capture prey, such as *Arthrobotrys* which snares nematodes on which it feeds. Many fungi are **parastitic**, feeding on living organisms without killing them. Ergot, corn smut, Dutch elm disease, and ringworm are all diseases caused by parasitic fungi.

Mycorrhizae are a symbiotic relationship between fungi and plants.

Most plants rely on a symbiotic fungus to aid them in acquiring water and nutrients from the soil. The specialized roots which the plants grow and the fungus which inhabits them are together known as **mycorrhizae**, or "fungal roots". The fungus, with its large surface area, is able to soak up water and nutrients over a large area and provide them to the plant. In return, the plant provides energy-rich sugars manufactured through photosynthesis. Examples of mycorrhizal fungi include truffles and *Auricularia*, the mushroom which flavors sweet-and-sour soup.

In some cases, such as the vanilla orchid and many other orchids, the young plant cannot establish itself at all without the aid of its fungal partner. In liverworts, mosses, lycophytes, ferns, conifers, and flowering plants, fungi form a symbiotic relationship with the plant. Because mycorrhizal associations are found in so many plants, it is thought that they may have been an essential element in the transition of plants onto the land.

SYSTEMATICS

There are also two conventional groups which are not recognized as formal taxonomic groups (ie. they are polyphyletic); these are the Deuteromycota (fungi imperfecti), and the lichens. The **Deuteromycota** includes all fungi which have lost the ability to reproduce sexually. As a result, it is not known for certain into which group they should be placed, and thus the Deuteromycota becomes a convenient place to dump them until someone gets around to working out their biology.

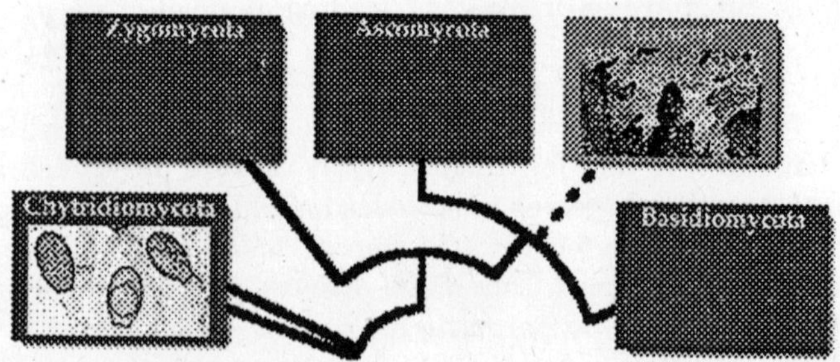

Fungi are usually classified in four divisions: the **Chytridiomycota** (chytrids), **Zygomycota** (bread molds), **Ascomycota** (yeasts and sac fungi), and the **Basidiomycota** (club fungi). Placement into a division is based on the way in which the fungus reproduces sexually. The shape and internal structure of the **sporangia**, which produce the spores, are the most useful character for identifying these various major groups.

Unlike other fungi, the **lichens** are not a single organism, but rather a symbiotic association between a fungus and an alga. The fungal member of the lichen is usually an ascomycete or basidiomycete, and the alga is usually a cyanobacterium or a chlorophyte (green alga). Often the fungal partner is unable to grow without the algal symbiont, making it difficult to classify these organisms. They will be treated here as a separate group, but it should be realized that they are neither single organisms, nor a monophyletic group.

It should also be noted that some organisms carry the name of mold or fungus, but are NOT classified in the Kingdom Fungi. These include the slime molds and water molds (Oomycota). The slime molds are now known to be a mixture of three or four unrelated groups, and the oomycetes are now classified in the Chromista, with the diatoms and brown algae.

MORE ON MORPHOLOGY

Like plants and animals, fungi are eukaryotic multicellular organisms. Unlike these other groups, however, fungi are composed of filaments called **hyphae**; their cells are long and thread-like and connected end-to-end, as you can see in the picture below. Because of this diffuse association of their cells, the body of the organism is given the special name **mycelium**, a term which is applied to the whole body of any fungus. When reproductive hyphae are produced, they form a large organized structure called a **sporocarp**, or mushroom. This is produced solely for the release of spores, and is not the living, growing portion of the fungus.

In addition to being filamentous, fungal cells often have multiple nuclei. In the chytrids and zygomycetes, the cells are **coenocytic**, with no distinction between individual cells. Rather, the filaments are long and tubular, with a cytoplasm lining and large vacuole in the center. By contrast, the ascomycetes and basidiomycetes are **septate**; their filaments are partitioned by cellular cross-walls called **septa**. The structure of these septa varies, and is taxonomically useful.

Another feature of fungi is the presence of **chitin** in their cell walls. This is a long carbohydrate polymer that also occurs in the exoskeletons of insects, spiders, and other arthropods. The chitin adds rigidity and structural support to the thin cells of the fungus, and makes fresh mushrooms crisp. Most members of the kingdom Fungi lack flagella; the structures are completely absent in all stages of their life cycle. The only exception are the chytrids, which produce flagellated gametes. The absence of flagella then, is a synapomorphy which unites all the remaining groups of fungi. This has had a tremendous impact on fungal biology, because it means that no fungus can produce motile gametes, and two organisms must therefore come into direct physical contact to effect sexual reproduction.

PARASEXUAL ANALYSIS IN TRICHODERMA REESEI USING PROTOPLAST FUSION

Protoplast fusion techniques were used to induce parasexuality in *Trichoderma reesei* and the products of fusion and segregation analyzed in order to get genetic data regarding the three cellulase markers, exoglucanase (exo), endoglucanase (endo) and beta-glucosidase (beta- glu) and their location with respect to other auxotrophic markers. Five mutants derived from *T. reesei* QM 9414, a hypercellulase producer, were used in this study.

Thirty-eight hour old conidia at a concentration of 1 x 10(6) spores/ml, after 1 h preincubation, were treated with 8 mg/ml Novozyme 234 for 5 h in medium containing 0.4 M (NH4)2SO4 as the osmoticum. Protoplast fusion was performed using polyethylene glycol (PEG) according to the procedures of Bos et al. (1983 Experientia Suppl. 45:298-299). Diploids were confirmed by their ability to segregate in the presence of p-fluorophenylalanine (pFPA). pFPA (7 g/ml) incorporated in complete medium (g/l: glucose, 10; yeast extract, 3; KH2PO4, 10;agar, 20) was used to induce marker segregation. The haploidizing medium was supplemented with amino acid requirements of both the parents to overcome selection against auxotrophic markers. The segregating sectors were picked randomly, purified and repeatedly subcultured on pFPA plates until stable sectors were obtained and then characterized on minimal medium (yeast extract was deleted from complete medium) supplemented with amino acid(s) as required (to a final concentration of 10 mM). Cellulase production, in complete medium broth containing 1% microcrystalline cellulose as the carbon source was measured in terms of exoglucanase, endoglucanase and beta-glucosidase activity (Sandhu and Kalra 1982 Trans. Brit. Mycol. Soc. 75:281-286). In a polygenic situation, as appears to be the case in the three cellulase components, the polygenes which represent an increase in the metric character above the mean value were taken as positive and those below it as negative.

Induced heterokaryosis via protoplast fusion was used to isolate seven fusants (Table 1) with different combinations of genetic markers. Auxotrophic and cellulolytic characterization showed that the segregants of each cross were distributed into various genotypic classes, both parental and non-parental, although the allelic ratios and the independent assortment of markers, as is common in haploid parasexual progeny (Croft and Dales 1981 Proc. Int. Symp. Basel pp. 179-186) did not follow the Mendelian expectation. The average recombination fractions as calculated from the seven crosses (Table 2) were found to be low among ade-lys, thr-ade and thr-lys, indicating absence of independent segregation and linkage between them. In parasexual progeny the fact that any recombinants are formed between apparently linked markers suggests that the rare event of mitotic crossing over has taken place (Bradshaw et al. 1983 J. Gen. Microbiol. 129:2117- 2123).

Table 1. Protoplast fusion between mutant strains of Trichoderma reesei QM9414

Cross	Parents	Auxotrophic Character	Cellulase Character
C	Tr 28	thr$^-$ lys$^-$	exo$^-$ endo$^+$ β-glu$^-$
	Tr 155/3-25	ade$^-$ lys$^-$	exo$^-$ endo$^-$ β-glu$^-$
D	Tr 28	thr$^-$ lys$^-$	exo$^-$ endo$^-$ β-glu$^+$
	Tr 48	thr$^-$ arg$^-$	exo$^+$ endo$^+$ β-glu$^+$
E	Tr 84	thr$^-$ leu$^-$	exo$^-$ endo$^-$ β-glu$^-$
	Tr 155/3-25	ade$^-$ lys$^-$	exo$^-$ endo$^-$ β-glu$^-$
F	Tr 48	thr$^-$ arg$^-$	exo$^+$ endo$^+$ β-glu$^+$
	Tr 155/3-25	ade$^-$ lys$^-$	exo$^-$ endo$^-$ β-glu$^-$
G	Tr 48	thr$^-$ arg$^-$	exo$^+$ endo$^+$ β-glu$^+$
	Tr 162	ade$^-$ lys$^-$	exo$^+$ endo$^+$ β-glu$^+$
H	Tr 48	thr$^-$ leu$^-$	exo$^-$ endo$^-$ β-glu$^-$
	Tr 28	thr$^-$ lys$^-$	exo$^-$ endo$^+$ β-glu$^+$
A	Tr 84	thr$^-$ leu$^-$	exo$^-$ endo$^-$ β-glu$^-$
	Tr 48	thr$^-$ arg$^-$	exo$^+$ endo$^+$ β-glu$^+$

Low recombination fractions show that either the markers are closely linked and crossing over frequent or that they are far apart and crossing over infrequent. High recombination between markers which are linked, as in the case of thr-ade in cross C, could be the result of mitotic crossing over having occurred prior to segregation giving rise to a clone of recombinants (Debets et al. 1989 Can. J. Microbiol. 35:982-988). Other auxotrophic markers registered high recombination fractions revealing the absence of linkage between them. Recombination data of the genes responsible for the formation of cellulolytic enzymes showed their presence on separate linkage groups. In *T. koningii*, Wey et al. (1991 Fungal Genetics

Newsl. 38:46) found gene probes of cellohydrolase and endoglucanase to hybridize to the same chromosome. As is evident in Table 2, the recombination fractions of the same marker pairs are close but not reproducible, because even slight variations in concentration of haploidizing agent are critical. One thr+ segregant was isolated from each of the two crosses made between two thr parents indicating the non-allelic nature of the mutations.

Table 2. Combined data of genetic recombination of auxotrophic and cellulolytic markers in seven crosses

Marker pair	Cross							Average recombination fraction
	C	D	E	F	G	B	A	
thr-arg				34.75 ± 5.05	47.36 ± 6.20			41.07 ± 5.98
arg-ade				34.75 ± 5.05	47.36 ± 6.20			41.07 ± 5.95
arg-lys				34.78 ± 5.05				34.78 ± 5.05
ade-lys				0	0	0		0
thr-ade	29.80 ± 5.05		13.50 ± 2.80	0	0			10.70 ± 5.13
thr-lys			13.50 ± 2.80	0	0			4.50 ± 6.33
lys-leu						64.04 ± 7.90		64.04 ± 7.90
thr-leu						86.56 ± 10.79		86.50 ± 10.79
leu-ade						78.26 ± 9.30		78.26 ± 9.30
arg-leu							20.42 ± 7.90	20.41 ± 7.90
thr-exo				45.60 ± 6.60	52.63 ± 6.90			49.11 ± 6.75
arg-exo		42.86 ± 9.20		54.35 ± 7.90	57.89 ± 7.60		75.55 ± 10.71	57.65 ± 8.85
ade-exo				45.60 ± 6.60	52.63 ± 6.90			49.11 ± 6.75
lys-exo		47.60 ± 10.26		45.60 ± 6.60				46.60 ± 5.40
leu-exo							16.32 ± 2.26	16.32 ± 2.26
thr-endo	63.15 ± 8.29			52.17 ± 7.60				57.66 ± 7.50
arg-endo		71.43 ± 15.47		56.57 ± 8.20			59.21 ± 5.38	62.37 ± 10.68
ade-endo	59.40 ± 7.20			52.17 ± 7.60				55.68 ± 7.20
lys-endo		38.30 ± 5.20		52.17 ± 7.60		40.03 ± 5.03		43.63 ± 6.24
leu-endo						50.55 ± 6.18	40.00 ± 5.05	45.08 ± 5.59
thr-β-glu	56.16 ± 7.36			76.00 ± 11.13				66.08 ± 9.25
arg-β-glu				55.52 ± 8.20			73.46 ± 10.67	64.29 ± 9.33
ade-β-glu	80.70 ± 10.62			76.00 ± 11.13				78.35 ± 10.88
lys-β-glu				76.00 ± 11.13		46.87 ± 5.79		61.45 ± 8.46
leu-β-glu						36.25 ± 6.37	34.75 ± 5.57	35.22 ± 4.47
exo-endo		47.62 ± 10.26		45.60 ± 5.97			32.60 ± 4.99	41.98 ± 5.92
exo-β-glu				56.50 ± 8.24			19.74 ± 5.38	33.35 ± 4.79
endo-β-glu	45.60 ± 5.97			43.48 ± 6.20		50.69 ± 6.70	32.85 ± 6.05	47.28 ± 6.35

On pooling the data from the crosses studied, it can be estimated that *T. reesei* has at least six linkage groups with the following markers: I (thr, ade, lys); II (arg); III (leu); IV (exo); V (endo); and VI (beta-glu). Pulse field electrophoretic separation of intact *T. reesei* chromosomes, for which no previous karyotype is available, by Gilly and Sands (1991 Biotechnol. Letts. 13:477-482) also registered the presence of six chromosomes. The results of our studies indicate that the protoplast manipulation technique is a powerful tool for the location of markers and the identification of linkage groups in industrially important fungi like Trichoderma which lack inherent sexual processes.

SOME OBSERVATIONS CONCERNING SP AND URE-2 IN NEUROSPORA

We are currently seeking rec-2 which is proximal to am on linkage group V and are gleaning incidental information on genes in this region and elsewhere. So far, the following is apparent: sp is recessive: each of 15 heterokaryons forced between his-3 K874; sp B132 and am1 32213 strains by growth on Vogel's + glycine medium had wild-type morphology.

The gene order sp, ure-2, am inferred from two point crosses is confirmed by three point data: FGSC 3809 (a; ure-2 D74) was crossed to F11089 (A; sp B132, am B501). 320 random spores were isolated, 79% germinated. A gene in FGSC 4299 interferes with Kolmark's urease test: While seeking to order sp, ure-2 and am, we set up a cross between FGSC 4299 (a; ure-2 47, am 32213) and T9043 (A; his-3 K874; sp B132) . Of 320 random spores , 92% germinated and were scored for urease activity by dabbing conidia onto paper soaked in urea and a pH indicator to detect ammonia production (Kolmark 1969, Mutation Res. 8:51-63). By this test, T9043 was urease positive and FGSC 4299 urease negative. However, only 23% of the progeny were urease positive, suggesting the segregation of a second mutation unlinked to ure-2. There is preliminary evidence that this additional locus could be linked to his-3.

VECTORS FOR EXPRESSION AND MODIFICATION OF CDNA SEQUENCES IN NEUROSPORA CRASSA.

The quinic acid inducible qa-2 promoter of Neurospora crassa has been used to express cloned genes by a number of different groups. However, most of the commonly available sources of this promoter require extensive sub-cloning and modification before they can be used as effective expression vectors. We report the construction of two plasmids that allow direct cloning and subsequent expression in N. crassa. Each plasmid also has a number of useful features to aid in detecting and modifying insert sequences.

The first of these plasmids, pMYX2 is a derivative of pGEMQa-2P, originally constructed by S. Kang and obtained from J. Kinsey (University of Kansas Medical Center). The qa-2 promoter region including the QA-1F binding sites and transcriptional start point were cloned as a 1 kilobase (kb) SacI-KpnI fragment into the corresponding restriction sites of the Promega plasmid pGEM3Zf(+). We inserted the trpC terminator and poly-A signal sequences of Aspergillus nidulans (Cullen et al. 1987. Gene 57:21-26) from plasmid pMP6 (M. Plamann, Texas A&M University) as a 0.7 kb BamHI fragment into the uniquc BamHI of pGEMQa-2P to form an intermediate plasmid, pMYX1. In order to provide a selectable marker for use in N. crassa the benomyl resistant -tubulin gene of pSV50 (Orbach et al. 1986. Mol. Cell. Biol. 6:2452-2461) was inserted into the unique SalI site of pMYX1 as a 2.6 kb SalI fragment to form pMYX2 (Figure 1a). The second of the plasmids reported here, pMYX10 is a derivative of pWF1, originally constructed by Fecke et al. (1993. Fungal Genetics Newsl. 40:34-35). These workers placed both the qa-2 promoter, including all the sequences included in pMYX2, and the qa-4 terminator and poly-A signals into the Pharmacia plasmid pT7T3 to form pWF1 (Fecke et al. 1993. Fungal Genetics Newsl. 40:34- 35). We inserted the hygromycin resistance cassette originally constructed by Staben et al. (1989. Fungal Genetics Newsl. 36:79-81) and modified to include a N. crassa cpc-1 promoter by M. Plamann from plasmid pMP6 as a 3.1 kb HindIII fragment into the unique HindIII site of pWF1 to form pMYX10 (Figure 1b). Insert orientations in both pMYX2 and pMYX10 were determined by restriction mapping.

Both pMYX2 and pMYX10 can be used as expression vectors by cloning the sequences of interest into the unique SmaI site immediately downstream of the qa-2 transcriptional start point in each plasmid. The insert should carry the appropriate translational control sequences necessary for successful expression in N. crassa but does not require the presence of any transcriptional signals. Therefore, these vectors are quite suitable for expression of cDNA encoded gene products. Integration of the recombinant plasmids into the N. crassa genome can be monitored by selecting transformants for the tightly linked resistance markers. The presence of the promoter and insert regions of the plasmids can be verified by either genomic Southern or PCR analysis. Both plasmids include unique bacteriophage T7 promoters immediately upstream of the qa-2 promoter region. Use of a primer specific to the T7 promoter and a single primer homologous to sequences within the insert allows the presence of at least a single, intact recombinant within the genome to be detected by PCR amplification of the qa-2 promoter/insert junction. In addition, both plasmids contain f1 filamentous phage origins, (+) orientation in pMYX2, (-) orientation in pMYX10, allowing site-directed mutagenesis of recombinants without requiring subcloning from the expression vector.

These vectors have been deposited in the FGSC. This work was supported by an SBIR Phase I award from NIAID to Mycotox, Inc.

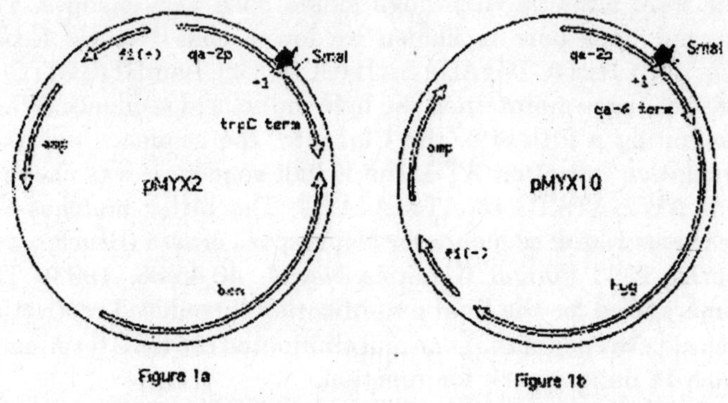

Figure 1: Map of pMYX2 (1a) and pMYX10 (1b). The qa-2 promoter region of each plasmid is indicated as qa-2p. The transcriptional start site is denoted as the large, filled arrow marked +1 on each map. The location of the unique SmaI cloning site is also indicated. The terminator regions are denoted as trpC term for pMYX2 and qa-4 term for pMYX10, indicating the Aspergillus nidulans trpC terminator and the N. crassa qa-4 terminator, respectively. The markers used for selection of transformants are indicated as ben (benomyl resistance) for pMYX2 and hyg (hygromycin resistance) for pMYX10. The f1 origins of replication and the ampicillin resistance genes of each plasmid are denoted as f1(+ or -) and as amp. The complete predicted sequence of pMYX2 is available.

IMPROVED VECTORS FOR SELECTING RESISTANCE TO HYGROMYCIN

Resistance to hygromycin B is an important dominant selectable marker in fungal transformation. Our goal was to improve vectors for hygromycin selection by making the gene more compact, by eliminating sites for commonly used restriction enzymes, and by subcloning the modified gene into convenient vectors. These improvements were made by modifying pCSN43 (Staben et al. 1989 Fungal Genetics Newsl. 36:79-81) through three rounds of megaprimer mutagenesis (Aiyar and Leis, 1993 Biotechniques 14:366-368), a technique based on polymerase chain reaction amplification. Plasmid pCSN43 has a 2.4 kb SalI fragment containing the bacterial hph gene (Gritz and Davies, 1983 Gene 25:179-188), encoding hygromycin B phosphotransferase, under control of the Aspergillus nidulans trpC promoter and terminator (Mullaney et al. 1985 MGG 199:37-45) (Fig. 1a). Four restriction enzyme sites in the hph gene were eliminated through single base pair changes. The changed base pair is shown by lower case letters: EcoRI (GAgTTC); PstI (CTGgAG); SstII (CCGCGc); BamHI (GGtTCC). These changes maintained the hph amino acid sequence. While eliminating a fifth site, the ClaI site, the sequence near the translation initiation ATG, the Kozak sequence, was changed from ATCGATATG to ATCcAaATG. The latter matches the consensus Kozak sequence for Neurospora crassa (Bruchez and Eberle, 1993 Fungal Genetics Newsl. 40:85-88, 1993). The primers used for the final amplification introduced restriction sites at both ends of the gene and eliminated the trpC terminator which is unnecessary for function.

The product from the final mutagenesis (Fig. 1b) was a 1.4 kb fragment with EcoRI, SalI and HpaI fragments at both ends. The HpaI sites were introduced to provide a convenient blunt end fragment for subcloning without the need for Klenow fill-in. The final product lacks the original BamHI, ClaI, EcoRI, PstI, and SstII sites as well as the trpC terminator.

The engineered hph gene was subcloned as follows. Plasmid pCB1003 was produced by ligating the final mutatgenesis product into the EcoRI site of pUC19. Plasmid pCB1004 was produced by ligating the SalI fragment of pCB1003 into the BclI site of pBC SK+. This was performed by half site fill-in, thereby destroying both restriction sites. Plasmid pCB1004 has chloramphenicol resistance, has a functional lacZ gene allowing blue/white screening, and has unique sites for all the restriction enzymes that cut in the "SK" polylinker. These plasmids transform Magnaporthe grisea strains 4091-5-8 and CP987 with the same efficiency as pCSN43 and confer the same level of resistance to hygromycin B.

B=BamH1 C=ClaI E=EcoRI N=NdeI P=PstI S=SalI T2=SstII H=HpaI

A PROTOCOL GUIDE FOR THE N. CRASSA YEAST ARTIFICIAL CHROMOSOME LIBRARY

A yeast artificial chromosome (YAC) library of Neurospora crassa strain 74-OR23-1A has been constructed. This library has been used to clone 750 kb of contiguous DNA sequences from the centromere region of linkage group VII (M. Centola and J. Carbon. 1994. Mol. Cell. Biol. 14:1510-1519). The purpose of this article is explicitly to outline procedures that have been developed for library screening and chromosome walking. The library was constructed in the YAC vector pYAC4 (Burke et al. 1987. Science 236:806-812). This vector contains Tetrahymena telomere, Saccharomyces cerevisiae CEN4, and ARS1 DNA sequences that specify in cis full telomeric, centromeric and replication functions in S. cerevisiae. YAC clones are maintained in approximately single copy as highly stable linear "artificial chromosomes". pYAC4 also encodes the yeast TRP1 and URA3 for selection of YAC clones in the yeast host strain AB1380 MATa ade2-1 can1-100 lys2-1 trp1 ura3 his5 {psi+} (Burke et al. 1987). A model of a YAC clone is shown in Fig. 1. The N. crassa YAC library contains 2204 clones, with an average insert size of 170 kb. Inserts from 40 clones have been characterized and range in size from 75-260 kb.

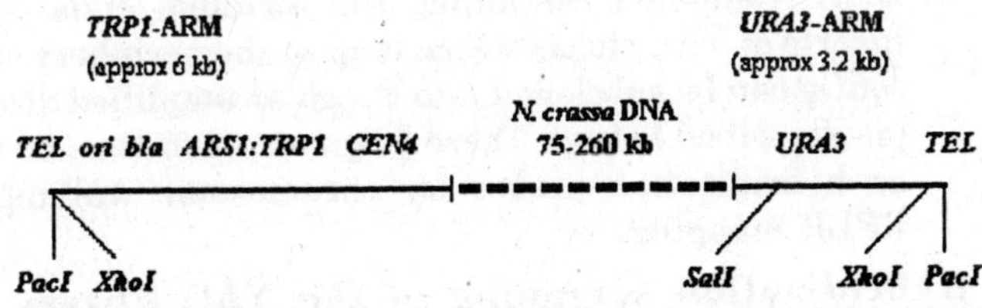

Figure 1. A schematic of a typical clone from the N. crassa YAC library is shown. The markers shown are described in the text.

The following wells in the library contain cultures that will not grow or grow slowly upon inoculation: 1:6A 1:11H 2:8F 2:11E 5:9G 6:5F 6:11E 10:1F 10:6A 10:9B 1 1:1H 11:2G 11:6A 12:7D 12:9C 14:3D 14:3E 14:9E 17:3G 17:6F 17:9H 17:10H 18:3E 19:7G 19:9F 20:2D 20:5E 20:7H 21:4F 22:3D 22:4A 22:9C 23:11D 24:4A 24:4B 24:4C. There are no clones in columns 5 to 12 on dish 24. Additionally, well 21:3E was contaminated with cells from well 21:3D. At least 5% of the clones besides those listed did not grow or grew poorly upon inoculation from a third generation copy of the YAC library distributed by the Fungal Genetics Stock Center (C. Yamashiro, unpublished observations).

Chromosome Walking

1. A protocol is outlined for hybridization screening of the YAC library. Positive YAC clones obtained from screening the library must be verified by Southern analysis. Mixed cultures within an individual well can occur; therefore, streak out cultures for single colonies on selective plates. Prepare genomic DNA from one or more individual colonies, and perform genomic Southern blot hybridizations with these DNAs as outlined below.
2. Several of the YAC clones obtained from a given library screening should be physically mapped with restriction endonucleases and the resultant maps compared. They should fit into a single set of contiguously overlapping sequences referred to as a "contig". Construction of a contig map assures the continuity of individual clones and avoids possible confusion resulting from multiple insert clones, whose restriction maps will not fit into the consensus maps generated from the bona fide majority of clones.
3. DNA fragments containing the terminal ends of the inserts of YAC clones which map to the periphery of the contig can be subcloned into E. coli or amplified directly (as described below). These fragments can then be used as hybridization probes for chromosome walking, or RFLP mapping.

Hybridization screening of the YAC library. (B. Brownstein et al. 1989. Science 244: 1348-1351, and B. Brownstein personal communication with modifications)

I. Preparation of a YAC library filter set.

1. Positively charged nylon membranes (MSI Inc, Westboro, MA) are layered onto the surface of large petri plates containing SD medium with casamino acids and adenine, "SD + CAA + A" (a supplemented minimal medium). Cultures from microtiter wells are replicated onto marked membranes using a multi-pin replicator. Each microtiter plate should be replicated onto a single membrane. Growth will take 3 to 4 days at 30oC. Fast growing colonies can be allowed to grow quite large (0.5 mm or larger) to allow the slow growing colonies to be well established (other protocols prefer smaller colonies so that cells are actively growing but lysis of old colonies works well here).
2. Prepare spheroplasting buffer CES/DTT/YLE (CDY). For 40 ml: Add 80 mg yeast lytic enzyme, 70,000 units/gram (ICN Biochemicals Irvine, CA) and 300 ml of 2 M dithiothreitol (DTT) to 40 ml CES solution (1 M sorbitol 100 mM Sodium citrate pH 7.0, 50 mM EDTA pH 8.0).
3. Place Whatman 3 MM or equivalent in lid of 150 mm Petri dish. Add approximately 7.0 ml CDY to Whatman and remove any bubbles (only a small amount of excess liquid should be present in lid). Carefully place a membrane containing YAC colonies onto the CDY saturated paper. Carefully remove bubbles from under the membrane.
4. Place the bottom of the Petri dish over the membrane and seal dish well with parafilm. Incubate 2 days 30-32oC.
5. Colony lysis and DNA denaturation is done by placing the membranes sequentially on Whatman 3MM paper saturated with the following solutions at room temperature for the times shown:
   a. 10% SDS/100 ug/ml proteinase K for 10 min.
   b. 0.5 N NaOH for 10 min.
   c. Three washes of 200 mM Tris pH 7.5/2X SSC for 5 min each.

6. Air dry the membranes on 3 MM paper for 2 h minimum (can go overnight). Bake for 2 h in vacuum oven at 80oC. Wrap membranes in plastic wrap (do not stack membranes, arrange them side to side). Store at -20oC.

II. Hybridization.

1. Prehybridize filters overnight at 65oC in a prehybridization (prehyb) solution of 1 M NaCl, 1% SDS and 10% dextran sulfate (Oncor Inc. Gaithersburg, MD). Four sets of six filters were each prehybridized with 50 ml of prehyb solution. Filters were incubated in plastic "Tupperware" containers with the lids on (hybridization bags could also be used). Discard the prehyb solution if cell debris is present (lots of cell debris will be present after the first prehybridization of the filter set).
2. Hybridization: Use 1 to 5 x 10^5 cpm of 32P-labeled probe/ml of prehyb solution. Boil carrier salmon or herring sperm ssDNA (100 ug/ml in final hybridization) with the probe for 10 min to reduce background caused by non- specific hybridization. Add labeled probe with carrier DNA to hyb cocktail in fresh Tupperware containers then add filters from prehybridization. Four sets of six filters were each hybridized with 50 ml of cocktail. Hybridize at 65oC (use from 42oC to 65oC depending on stringency) overnight with shaking. To save money, we hybridized two sets of six filters one night, then reused the hybridization cocktail for the remaining filters the following day.

Note: YAC vector pYAC4 contains the entire pBR322 DNA sequence. Do not use probes with this sequence, otherwise your probe will hybridize to all the YAC clones. Alternatively, a 50-fold molar excess of cold competitor pBR322 DNA can be added to the hybridization reaction to reduce background. Similarly, if labeled cosmid DNA is used as a probe use a 50-fold molar excess of cold cosmid vector (i.e., lacking insert DNA) to decrease background (vector-vector) hybridization.

3. Remove membranes for washes on a gyratory shaker. Wash times should be adjusted to yield background signal acceptable to you.

a. Wash in 2 changes of 2X SSC/ 1% SDS for 30 min each at 65oC.

b. Wash in 2 changes of 0.5X SSC/ 1% SDS for 30 min at 65oC. For higher stringency use 0.1X SSC/1 % SDS for 1 h at 75oC.

4. Wrap membranes in plastic wrap and expose to film.
5. Rehybridization: We did not strip our library filters. Instead, we prepared two filter sets and each set was allowed to decay approximately 3 months prior to rehybridization. If filters are to be stripped do not allow them to dry out at any time (store wet at -20oC).

Selective medium (SD + CAA +A) 0.67% Bacto-yeast nitrogen base without amino acids, 2% dextrose, 0.5% casamino acids, 15 mg adenine/liter. For solid medium add 2% agar.

YAC Restriction Mapping and Contig Building

Restriction maps of YAC clones can be generated quickly by partial digestion and indirect end labeling (M. Centola and J. Carbon. 1994. Mol. Cell. Biol. 14:1510-1519) as follows: Chromosomal DNAs of the YAC clones imbedded in agarose "DNA plugs" are prepared. DNA plugs are partially digested with a restriction endonuclease which recognizes an 8 bp DNA sequence.

Partial digestion reactions are fractionated on CHEF gels and blotted to membranes. Membranes are then hybridized with probes specific for one of the YAC vector arms. A series of positive hybridization bands of increasing size will be imaged upon autoradiography of the membrane in each lane containing a partial digestion reaction. The size of each band is calculated relative to size standards run on the fractionation gel. The difference in size between sequential bands is equal to the distance between restriction sites in a given clone. Resultant restriction maps can be verified by rehybridizing stripped blots with a probe specific for the opposite arm of the YAC clone. Restriction maps of individual clones can then be overlapped and contigs can be generated.

I. Preparation of chromosomal DNA plugs of YAC clones (Kuspa et al.. 1989. Proc. Natl. Acad. Sci. USA 86:8917-8921).

A. Small Scale Preparation:

1. Grow 3 ml culture of yeast strain in SD+CAA+A medium at 30oC to stationary phase (2-3 days).
2. Pellet 1.5 ml of cell culture in a microfuge tube and wash pellet with 1 ml 0.5 M EDTA pH 8.0.
3. Remove EDTA solution, respin cells and remove ALL remaining EDTA solution with a pipetman.
4. Gently resuspend cells in 10.8 ul spheroplasting solution.
5. Incubate at room temperature (RT) for 5 min.
6. Add 25.2 ul of molten 1.2% LMP agarose in SCE cooled to 45oC (final concentration of agarose is 0.6%).
7. Pour into plug molds. Solidify at RT for approximately 20 min.
8. Move plugs into microfuge tube. Add overlay solution to fill tube. Incubate overnight at 37oC to produce spheroplasts.
9. Remove overlay solution carefully.
10. Add 1 ml of NDS; incubate for 2 days at 50°C.

B. Large Scale Preparation

1. Grow 50 ml culture of yeast strain in a 125 ml flask, with shaking, to stationary phase (2-3 days) at 30°C in SD+CAA+A medium; cultures will turn pink.
2. Pellet cells 5 min at approximately 1000 x g.
3. Wash with 100 ml 50 mM EDTA pH 8.0 (EDTA solution).
4. Remove EDTA solution, respin cells and with a pipetman remove ALL remaining EDTA solution.
5. Gently resuspend pellet in 400 ul spheroplast solution.
6. Incubate at RT 5 min: briefly warm cells to 45°C (10-20 sec).
7. Add 800 ul 1.2% LMP agarose in SCE. Cool molten agarose to 45-48oC prior to addition. Mix gently and completely. Quickly pour into plug molds. Solidify at RT for approximately 20 min.
8. Move plugs into 15 ml Corex tube add 12 ml overlay solution. Incubate overnight at 37°C to produce spheroplasts.

9. Remove overlay solution carefully.
10. Add 10 ml of NDS and incubate overnight (for 1-2 days) at 50°C. Plugs can be stored in NDS at 4°C.

Solutions with volumes for large scale preps are given below. Final volumes can be scaled up or down.

Spheroplast solution

1 ml SCE

25 ul 2-mercaptoethanol

1 mg Zymolyase (100T) (ICN Biochemical, CA)

1.2% LMP agarose

0.12 g LMP agarose

10 ml SCE (melt agarose)

Overlay solution

46 ml 0.5 M EDTA

0.25 ml 2M Tris pH 8.0

3.75 ml 2-mercaptoethanol

NDS

1 mg/ml Proteinase K

1% N-lauroyl sarcosine in 0.5 M EDTA pH 9.0

SCE

1 M sorbitol

0.1 M sodium citrate, pH 5.8

10 mM EDTA (filter sterilize)

Plug molds: Use 1 cc syringe barrels with the ends sealed with parafilm or slide holders that come with common microscope slides. Tape up sides of slide holder and pour in molten agarose to form plugs that fit nicely into gel wells.

II. Partial Restriction Enzyme Digestion of YAC DNA Plugs

Plugs must be extensively washed prior to digestion. Wash a given volume of plugs 5 times with 25 volumes of 10 mM Tris pH 8.0, 50 mM EDTA at RT for 2 hr/wash. Plugs can be washed and then stored in wash solution at 4oC for months prior to digestion.

1. Add the equivalent of approximately 20 ul volume of solid chromosomal DNA plug/tube to two microfuge tubes (approximately 0.2-1 ug of DNA). Two digestion reactions will be done on each clone.
2. To each tube add 100 ul 0.5 mM phenylmethylsulfonyl fluoride (PMSF; USB, Cleveland, OH) in TE buffer. Incubate reactions at 37°C for 30 min. PMSF is highly unstable in aqueous solution (half life = 30 min); prepare a fresh 100 mM stock in 100% ethanol and dilute to 0.5 mM immediately before each use.
3. Remove PMSF wash solution and add 1 ml of TE buffer to plug. Incubate at 37°C for 30 min and repeat wash once. Remove final wash solution completely.
4. Add 1/10 volume of restriction enzyme buffer (10 x concentrate) to plug and melt mixture at 68°C for 5 min. Plugs should be completely molten.
5. Cool plugs to 37°C and add 0.2 and 1.0 units of pre-warmed (37oC) restriction enzyme to each of the two reactions, respectively. The plug should still be molten. Mix gently with a pipet tip. Incubate at 37°C for 30 min.
6. Add 1 ul of 0.5 M EDTA (pH 8.0) and remelt plugs at 68°C for 5 min and the molten reactions loaded onto an agarose gel (see below).

Note: Complete restriction enzyme digests of the DNA within agarose plugs can be done as described above if a large amount of enzyme is used (we typically use 12-30 units of enzyme/20 ul of plug) and incubate reactions at 37°C for 3 hr or more.

III. Using CHEF Gel Analysis to Resolve YAC Clones

Partial digests are fractionated on a 1% agarose gel in 0.5 x TBE (1 x TBE = 90 mM Tris-borate, 2 mM EDTA). Gels (10 x 10 cm) are subjected to electrophoresis at 6.0 V/cm for 15 h in a CHEF gel apparatus (BioRad, Richmond CA), using a 0.2 to 13 sec pulse ramp, with a 120o pulse angle, at 14oC. Lambda concatemers and a 5 kb ladder (pBR328 concatemer, BioRad) size standards are also loaded on the gel. Stain gel with ethidium bromide, and photograph with a ruler atop the gel so migration distances of standards can be determined. These conditions will resolve DNA fragments ranging from 5-250 kb and are thus suitable for both partial digest mapping and for analysis of complete digestion products by Southern hybridization.

Some YAC clones are >250 kb. Resolution of uncut YAC clones can be done using the following conditions: 1% agarose gels prepared and subjected to electrophoresis in 0.5 x TBE ffer at 6.0 V/cm, using a 6-12 sec ramped switch time, at 12oC for 16-20 hr. These conditions will resolve clones up to 400 kb in size. Plugs can be placed into the wells of a CHEF gel by balancing the plug on a glass coverslip and pushing the plug into the well with a pipet tip. Alternatively, as suggested by Virginia Pollard, plugs can be placed onto the teeth of the well comb and the agarose gel cast with the comb in place. One needs to check for any dislodged plugs which can be repositioned using a Pasteur pipet.

IV. Southern Hybridization

Southern blotting of high molecular weight DNA requires extensive nicking of the DNA prior to transfer, and a large volume of transfer buffer to be used during capillary transfer.

1. Nick high molecular weight DNA within the ethidium bromide stained gel using a commercial UV cross linker. Use settings recommended by the vendor; BioRad crosslinkers have a program for nicking pulse field gels that works nicely. Alternatively, place the gel into 0.2 N HCl for 30 min at RT with gentle agitation.
2. Denature gel in 0.5 M NaOH, 1.5 M NaCl for 30 min.
3. Neutralize in 1 M Tris pH 7.5, 1.5 M NaCl for 30 min.
4. Transfer to a positively charged nylon membrane using standard capillary blotting procedures with the following modifications: a. Use a very large wick of 3MM paper, approximately 35 cm long x 30 cm wide, under the gel. b. Use 2 liters of 10x SSC for each transfer. Use a large volume of absorbent material above the gel. Use two glass plates (approximately 30 cm x 20 cm) as weights above the absorbent, excessive weight will compress the gel and prematurely stop the transfer.
5. Standard Southern hybridization conditions can be used.
6. Hybridization probes specific for the left and right arms of the pYAC4 vector can be obtained from the following

sources: a. Gel-purified 1.7 kb BamHI-PvuII fragment from pBR322 hybridizes specifically to the URA3 - encoding YAC vector arm (URA3-arm). b. Intact pBluescript (Stratagene, La Jolla, CA) hybridizes specifically to the TRP1- encoding YAC vector arm (TRP1-arm).

Isolation of Terminal Restriction Fragments from Cloned DNA Inserts in YAC Clones

Terminal restriction fragments of the insert DNA ("terminal fragments") from the TRP1-arm can be directly subcloned into E. coli by plasmid rescue and terminal fragments from the URA3-arm can be subcloned into E. coli. Alternatively, terminal restriction fragments from either end of YAC insert DNA can be cloned by ligation-mediated PCR. Although large regions of DNA (up to 30 kb) can be subcloned by plasmid rescue, the procedures are involved and at times frustrating. We therefore recommend the use of ligation-mediated PCR for obtaining small probes for chromosomal walking.

To isolate terminal end fragments of YAC inserts by plasmid rescue: Digest total genomic DNA from a given YAC clone with XhoI, or PacI (a double digest with XhoI and SalI should also work). These enzymes cut the vector arm once on the telomere proximal side of the E. coli bla and ori. These enzymes should also cut somewhere in the insert DNA. The digestion reaction is then diluted and ligated to circularize the DNA fragment containing bla, ori and the terminal restriction fragment of the insert DNA. A portion of the ligation reaction is electroporated into E. coli and AmpR colonies selected and subsequently screened for the rescued plasmid. Very large (>25 kb) plasmids do not rescue well into E. coli; therefore, determine the expected size of the rescue plasmid for a given enzyme before doing a rescue by preparing a Southern blot of CHEF gel fractionated genomic DNA cut with PacI, XhoI or XhoI/SalI from a given YAC clone and probing with a TRP1-arm specific probe. This will yield the approximate size of the terminal restriction fragment to be rescued. This information should be used to ensure that the rescued plasmid is of the correct size.

Similarly for subcloning terminal fragments from the URA3-arm, digest total genomic DNA from a given YAC clone with

PacI, XhoI or XhoI/SalI and shotgun clone the entire genome into an E. coli vector (we used pBluescript). The yeast URA3 gene heterologously rescues the Ura phenotype of E. coli strain DB6656 lacZ 624(Am) trp-49(Am) pyrF79::Mu rpsl179 hsdR27 (Bach, et al. 1979 Proc. Natl. Acad. Sci. USA 76:386). The ligation reaction is introduced into this strain by electroporation and Ura+ transformants selected.

I. Preparation of miniprep DNA from liquid cultures of YAC clones (L. Clarke and M. Baum, personal communication).

1. Inoculate 10 ml of SD + CAA + A with yeast cells late in the day in a 125 ml or larger flask. Incubate overnight with shaking at 32oC until culture reaches an OD660 of 0.6-0.7 (higher concentration cultures yield DNA resistant to restriction enzyme digestion). Cell cultures can be transferred to 50 ml disposable corning tubes upon reaching the proper OD, pelleted and stored at -20oC for several days before beginning DNA preps.
2. Wash with 2 ml 1 M sorbitol .
3. Loosen cell pellets before addition of SCE by brief vortexing, then resuspend cells gently in 2 ml SCE containing 0.2 mg Zymolyase 100T and 5 ul 2-mercaptoethanol. Incubate at 37oC for 1 h (never vortex the cells once Zymolyase is present).
4. Centrifuge cells at 800 x g for 4 min, and wash pellet with 2 ml SCE.
5. Remove wash, and resuspend cells in 1 ml NE solution (0.15 M NaCl, 0.1 M EDTA pH 8.0) containing 1% SDS. Add 5 ul 10 mg/ml DNAse free RNase, incubate 15 min at 37oC.
6. Add 5 ul proteinase K (20 mg/ml) and incubate another 15 min at 37o
7. Extract with 1.5 ml phenol/CHCl3/isoamyl alcohol (25:24:1). Invert gently 10 times. Lay tubes on their sides for 5 min. Spin 10 min at 1600 g in clinical centrifuge. Remove aqueous layer with P1000 pipetman using pipet tips with cut off ends (cut off ends so orifice is approximately 0.3 cm in width). Interface is prone to

mixing with aqueous phase so use caution. Repeat extraction once.

8. To aqueous layer add 0.12 ml of 7.5 M NH4OAc and 1.2 ml isopropanol, invert to mix. Spin at 8500 rpm for 10 min.
9. Wash pellet with 5 ml of 70% ethanol. Air dry pellet, and resuspend in 50 ul TE buffer.

   20 ml SCE: 10 ml 2 M sorbitol , 2 ml 1 M Sodium citrate pH 5.8, 2.4 ml 0.5 M EDTA pH 8.0, 5.6 ml H20.

Note: Be aware of the shear forces generated during the resuspension steps. Be gentle but thorough. Use cut off pipet tips when aliquoting DNA.

II. Plasmid rescue of the terminal restriction fragment from the TRP1-arm (L. Clarke and M. Baum, personal communication with modifications).

1. Digest 5 ul of miniprep DNA in a total reaction volume of 50 ul with 20-30 units of restriction enzyme for 3 h at 37oC. Miniprep yields will vary - 5 ul will yield about 2-5 ug. (save an aliquot of the digest to run on a pulse-field gel to confirm complete digestion and expected size as described above).
2. Add 40 ul H_2O + 10 ul 3 M NaOAc + 40 ul phenol/CHCl3/ isoamyl alcohol (25:24:1), "PCI" and mix.
3. Microfuge for 5 min to separate phases and transfer aqueous phase to a new tube.
4. Precipitate with 2 volumes of ethanol at -20oC for 1.5 h.
5. Spin in microfuge at full speed for 20 min.
6. Wash pellet with 0.5 ml 70% ethanol.
7. Resuspend pellet in 650 ul TE buffer and reprecipitate with 100 ul 7.5 M NH4OAc + one volume of isopropanol (second precipitation helps to further clean up the reaction).
8. Pellet and wash as above.
9. Resuspend pellet in 1 ml of 1 x T4 DNA ligase buffer and add 2 units of T4 DNA ligase, incubate at room temperature overnight.

10. Split the reaction into two 500 ul aliquots; to each add 50 ul 3M NaOAc + two volumes of ethanol.
11. Store one aliquot at -20oC (back-up reaction) and the other precipitate at -70oC for 20 min.
12. Pellet and wash as before.
13. Resuspend pellet in 25 ul TE buffer.
14. Introduce a 1 ul aliquot of the reaction mix into E. coli by electroporation and select for AmpR colonies. Note: Plasmid rescues containing repeats will tend to delete if ampicillin concentration in the selective media is >35 mg/l. Standard transformations can be used; however, it is more difficult to get rescues with lower transformation efficiencies.

III. Subcloning of terminal restriction fragment from the URA3 arm.

1. Digest 1 ug of miniprep yeast DNA with 20-30 units of restriction enzyme in a 50 ul final volume for 3 hr at 37oC as described above.
2. Digest 1 ug of pBluescript (or appropriate vector) with 20 units of enzyme as in step 1.
3. Add 0.5 unit of calf intestinal alkaline phosphatase (Boehringer Mannheim, Indianapolis, IN) to the pBluescript reaction, incubate an additional 20 min at 37oC.
4. Extract each reaction with an equal volume of PCI.
5. Precipitate DNAs with 1/10 volume of 7.5 M NH4OAc + 1 volume of isopropanol.
6. Resuspend each reaction in 10 ul TE buffer, and check the DNA concentration by gel electrophoresis. (Alternatively, if you feel lucky assume your yield and proceed).
7. Ligate 100 ng of digested genomic DNA and 100 ng of prepared vector in a 20 ul reaction volume with 1-2 units of T4 DNA ligase overnight at 16oC.
8. Introduce 1 ul of the ligation reaction into E. coli strain DB6656 by electroporation and select for Ura+ transformants on E. coli minimal medium supplemented

with 50 ug/ml tryptophan, and 50 ug/ml ampicillin (T. Maniatis et al. 1982. Molecular cloning, N.Y.). Colonies will appear after 2-3 days.

IV. Ligation-mediated PCR.

Ligation-mediated PCR (LM-PCR; Mueller and Wold 1989. Science 246: 780-786) is a procedure to isolate DNA fragments for which the sequence of the DNA is unknown. This technique has been adapted for generating DNA fragments corresponding to the end of inserts within YAC clones (Kere et al. 1992 Genomics 14: 241-248). Below we describe the LM-PCR technique used for generating end fragments of inserts within YAC clones with several minor modifications.

1. Six oligonucleotides are needed (5' to 3'):
   L primer further from the cloning site (oligo L1)
   CACCCGTTCTCGGAGCACTGTCCGACCGC
   R Primer further from the cloning site (oligo R1)
   ATATAGGCGCCAGCAACCGCACCTGTGGCG
   L primer nearer the cloning site (oligo L2)
   TCTCGGTAGCCAAGTTGGTTTAAGG
   R primer nearer the cloning site (oligo R2)
   GTCGAACGCCCGATCTCAAGATTAC
   Linker long strand and linker primer
   GCGGTGACCCGGGAGATCTGAATTC
   Linker short strand
   GAATTCAGATC
2. Digest 1 ug of yeast DNA embedded in agarose plugs (approximately 20 ul of plug) to completion as described above using restriction enzymes leaving blunt ends, such as RsaI, AluI, PvuII, EcoRV, or ScaI (usually set up 2-4 different digests for each clone). One can also digest 1 ug of total genomic DNA obtained from the miniprep procedure (above) in a 20 ul reaction volume.
3. Use 5 ul of the restriction digest and ligate the linker by adding ligation buffer, 25 pmol linker, and 1-2 units of T4 DNA ligase in a total reaction volume of 20 ul. Leave at room temperature for >1 h.

4. Set up the PCR using 2-5 ul ligation reaction, 20 pmol long vector-specific primer (L1 or R1) and 10 pmol of linker primer in a total volume of 50 ul. Cycling conditions are 94oC for 1 min, 65oC for 2 min, and 72oC for 2 min for 30-35 cycles.
5. Visualize products by subjecting 5-10 ul to electrophoresis on a 1.5% agarose gel.
6. To make probes from these fragments (they still have significant amount of vector sequence at this time), use PCR amplification of the purified first PCR product with 20 pmol of short vector-specific primer (L2 or R2) and 10 pmol of linker primer in a total volume of 50 ul with the same cycle conditions used in step 4. Purifying the first PCR product is simple by first coring out a small amount of agarose containing the band of interest with a pasteur pipet and transferring to a microfuge tube. Add 100 ul of TE buffer and boil for 5 min. Use 1 ul for the second LM-PCR reaction. The PCR reaction can be done in the presence of 32P-nucleotide for a radioactively labeled probe or in the presence of digoxigenin labeled dUTP (Boehringer Mannheim) for a non-radioactive probe (the latter probe can be used according to protocols provided by Boehringer Mannheim with the Genius kits and systems for non-radioactive detection of nucleic acids).

2

Transformation of N. Crassa with a YAC Clone

I. INTRODUCTION OF A N. CRASSA SELECTABLE MARKER INTO A YAC CLONE

The pYAC4 vector does not encode a N. crassa selectable marker. Therefore, transformation of a given YAC clone is limited to clones that encompass selectable markers and must be carried out in appropriate mutant strains. Introduction of a clone from the library into N. crassa has been completed by transforming a qa-2, arom-9, inl strain of N. crassa (FGSC #3952) with total genomic DNA from YAC clone AB1380/YAC 12-10-H (M. Centola unpublished observations). This clone encompasses the qa-2+ gene. Standard spheroplast transformation procedures were used (Vollmer and Yanofsky 1986. Proc. Natl. Acad. Sci. USA 83:4869-4873) and 10 transformations using 1 ug of DNA/150 ul of competent cells yielded a single transformant.

To facilitate introduction of any YAC clone in the library into N. crassa a yeast integration vector carrying a dominant N. crassa selectable marker was constructed. The plasmid pLUShph (diagrammed in figure 2A) encodes the hygromycin phosphotransferase gene (hph) under the control of a modified cpc-1 regulatory region (Royer and Yamashiro 1992. Fungal Genetics Newsl. 39:76-79). This plasmid (pLUShph) also contains the yeast LYS2 gene, and an E. coli KanR gene and ori. This plasmid was constructed by inserting the 3.0 kb HindIII fragment from pMP6 into the unique HindIII site on pLUS (Hermanson et al 1991. Nucl. Acids Res. 18:4943-4948).

INTEGRATION OF THE PLUSHPH VECTOR INTO THE YAC CLONES

This plasmid can be site-direct integrated into the URA3-arm of the YAC clones by digesting pLUShph at the unique SalI site and introducing the linearized plasmid into the yeast hc

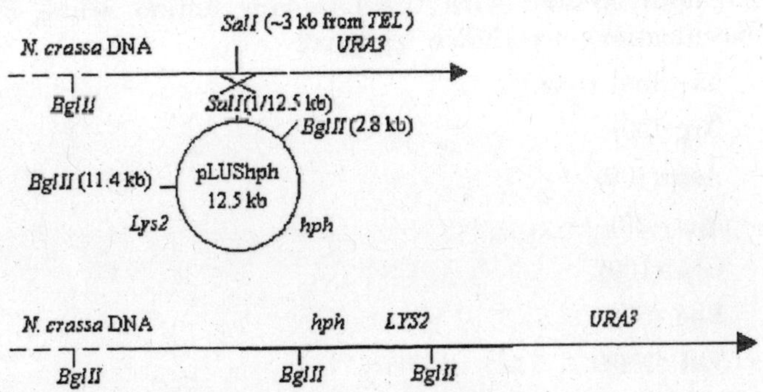

Figure 2. YAC integration vector pLUShph. A) A schematic of the YAC integration vector is shown. The region designated sup4-URA3 is homologous to a portion of the URA3-arm of the YAC clones. As described in the text, this region can be utilized to mediate site-specific integration of this vector into the URA3-arm of any YAC clone. B) A schematic of the integration event is shown. The product of integration contains the yeast LYS2 gene for phenotypic selection of integrated clones. A typical YAC integrant, shown at the bottom of the figure, also carries the N. crassa dominant selectable marker hph that can be used for selection of the YAC integrants upon introduction into N. crassa.

We have integrated this plasmid into several YAC clones by transforming competent yeast spheroplasts. However, lithium acetate transformations (Elble 1992. BioTechniques 13:18-20) are much easier to perform and transformation of LiOAc competent yeast cells with 1 ug of pLUShph linearized with SalI should also be successful. The integrants are selected on dropout media lacking lysine and tryptophan (see below). The DNA from the colonies displaying the Lys+ phenotype was prepared in agarose plugs.

Dropout medium - SD + Adenine + essential amino acids (no lysine)

0.67% Bacto-yeast nitrogen base without amino acids

2% Dextrose

15 mg adenine/liter

2% agar.

Supplemented with the following amino acids. Final concentrations are shown in ug/ml.

aa (final conc.)

Arg (20)

Asp (100)

Met (20)

Glu (100)

Phe (50)

Val (150)

His (20)

Ser (375)

Tyr (30)

Leu (60)

Thr (200)

II. Verification of Integration by Southern hybridization and Restriction Enzyme Analysis.

A. Detection of an integrated copy of the hph gene within YAC clones by Southern analysis

1. Prepare DNA from the Lys+ yeast colonies in agarose plugs (see above for plug preparation protocol).

2. Fractionate the undigested DNA from the Lys+ transformants and parental host strains on a CHEF gel (use conditions described above for 5-250 kb resolution).

3. Transfer the DNA to a nitrocellulose membrane.

4. Hybridize the membrane with a radioactively-labeled DNA fragment containing the hph gene. We used the 1.2 kb BamHI/ClaI DNA restriction fragment from pCSN44 that encompasses the hph coding region (available from the FGSC). YAC clones in which pLUShph has integrated will hybridize to the probe.

RESTRICTION MAPPING OF HPH CONTAINING YAC CLONES

1. Digest the DNAs from Lys+ YAC clones that hybridize to the DNA fragment containing the hph gene and the parental host strains with a restriction endonuclease which cuts inside pLUShph, but not inside pYAC4 (see "Restriction enzyme digestion of YAC DNA plugs" protocol shown above). We used the restriction endonuclease BglII.
2. Fractionate the digested DNAs on a CHEF gel (see "Using CHEF gel analysis to resolve YAC clones" protocol).
3. Transfer the DNA to a nitrocellulose membrane.
4. Probe the Southern blot with a radioactively-labeled DNA fragment which has homology with the YAC arm and pLUShph. We probed the Southern blot with the 1.7 kb BamHI/PvuII fragment from pBR322.
5. Expose the blot to film.
6. From the autoradiograph, determine the distance from the end of the telomere to the first restriction site within the insert of the parental YAC clone.
7. From the known restriction map of pLUShph, determine the size of the restriction fragments one would get if the plasmid were integrated at the unique SalI site. This is done by adding the distance from the end of the telomere to the SalI site on the YAC arm into which the plasmid was integrated to the distance between the SalI site on pLUShph and the first restriction site on the SUP4 side of the plasmid. Conversely, one will see a band with a size corresponding to the distance from the restriction site within the insert to the SalI site on pLUShph to the next restriction site on the URA3 side of pLUShph.

SEQUENCE AND ANALYSIS OF GENOMIC SEQUENCES UPSTREAM OF MEI-3.

The Neurospora crassa mei-3 mutation causes sensitivity to various DNA damaging agents (Newmeyer and Galeazzi 1977 Genetics 85:461-487). We have recently cloned and mapped a genomic fragment capable of transforming mei-3 spheroplasts to wild type (Cheng et al. 1993 Mut. Res. 294:223-234). These experiments predicted the putative coding sequence of mei-3 or at least the region encompassing the mei-3 mutation. It was determined by homology that Mei-3 belongs to the RecA-like group of proteins (Bishop et al. 1992 Cell 69:439-456) which are intimately involved in the recombination process and had been previously identified only in prokaryotes.

Since the identification of RecA-like proteins in Saccharomyces (Shinohara et al. 1992 Cell 69:457-470) and in Neurospora, our hypothesis that this important group of proteins may be highly conserved in other eukaryotes has been substantiated with the cloning of several Rec-A like proteins from mouse, chicken, lily, and human (Shinohara et al. 1993 Nat. Genet. 4:239-243). Using these data there is evidence that additional homology, upstream of our putative start site, to these other proteins exists (an additional 67% over 39 amino acids between Mei-3 and mouse Rad51: Figure 1). However, this region of homology lacks a start site or obvious splice sites. This might suggest the presence of an unidentified upstream start sequence and another exon without obvious splice sequences. In an attempt to address this possibility, we sequenced both strands of the previously unpublished genomic sequence, from -2519 to -271 bp upstream of the putative 5' end of mei-3.

The sequencing was accomplished by sequencing pBRC4 plasmid DNA, using custom made primers, and the Taq cycle sequencing system (United States Biochemical, Cleveland OH). Sequence analysis was performed using the Genetics Computer Group (GCG) programs. The codon preference table used in the analysis was provided by Dr. Mary Anne Nelson, Department of Biology, University of New Mexico.

The 2248 bp of DNA sequence from -2519 to -271 upstream of the putative start site of mei-3 on pBRC4 has been added to the Genbank depository under ascession number L02428 (March 1994). Comparisions made between the upstream region with yeast rad51, rad57, dmc1, with the mouse rad51 homolog, with the chicken rad51 homolog, and with the lily rad51 homolog did not reveal additional regions of homology between our upstream sequence and these proteins. Analysis of the upstream sequence did not reveal obvious consensus splice sites to tie any upstream open reading frames, or any regions with high coding preference, to the open reading frame we had originally identified as mei-3.

By homology we were unable to determine the location or existence of the putative upstream start site or exon possibly because the amino terminus ends of the eukaryotic RecA-like proteins tend to be divergent and/or mei-3 contains unusual splice sequences. Additional work is required to conclusively determine the 5' end of mei-3.

PRODUCTION OF TYROSINASE DEFECTIVE MUTANTS OF NEUROSPORA CRASSA

We have produced mutants defective in tyrosinase activity using the RIP procedure (Selker et al. 1987 Cell 51:741-752, Marathe et al. 1990 Mol. Cell. Biol. 10:2638-2644). The tyrosinase gene has been shown to be a useful reporter in gene expression experiments (Kothe et al. 1993 FGN 40: 43-45). The availability of tyrosinase mutants will assist in these studies and will also be of use for future protein engineering experiments on tyrosinase (Fuentes et al. 1993 FGN 40:38-39).

A qa-2 arom-9 al-2 strain of N. crassa was cotransformed with the qa-2-containing plasmid pRAL-1 (Akins and Lambowitz 1987, Cell 50:331-345) and the plasmid GRG-1/TYR103 (Kothe et al. 1993 FGN 40:43-45), which contains the tyrosinase gene sequences. The transformed strain was crossed with the wild-type strain 74-OR23-1A. Mutants defective for tyrosinase activity were isolated and used as males in crosses with am132 mutants of both mating types. Several progeny containing both mutations (am132and T-) were isolated and analysed by genomic Southern blotting.

A comparison of the progeny DNA with the wild-type DNA shows that isolate 32C (an isolate from the mating of a transformant with 74-OR23-1A) has an extra copy of the T gene as demonstrated by the SalI and BamHI digestion patterns. The T gene hybridizing sequence in isolate 32CxA15 (a T-mutant isolate from the mating of 32C with the am mutant) has lost the ~450 bp SalI fragment. This has probably arisen due to the methylation of the internal cytosine residue of the SalI recognition site which prevents its cleavage. The differential restriction pattern observed for the isochizomeric enzyme pair MspI/HpaII in this strain also indicates the presence of cytosine methylation. These methylation events are often accompanied by the base pair changes associated with the RIP phenomenon (Cambareri et al 1989 Science 244:1571-1575).

From a number of sexual crosses (>100) it is clear that all of our tyrosinase defective isolates are female sterile. The T- mutants are unable to complete protoperithecial development. The isolates can, however, be used as the male partner in genetic crosses. The regulatory mutant ty-1 which is defective in tyrosinase induction is also female sterile. However, unlike the ty-1 mutants which have a velvet morphology the T- mutants exhibit a normal growth phenotype. The segregation of the tyrosinase defective phenotype can be followed either by directly assaying for tyrosinase activity or by following the female sterility phenotype.

Isolates with mutant tyrosinase genes (am132 T- al-2 and T- al-2 mutants) have been deposited with the FGSC, which will be of value to investigators using the tyrosinase gene as a reporter in transformation experiments.

TERGITOL ENABLES THE RAPID AND INEXPENSIVE SCORING OF NUTRITIONAL AND DRUG-RESISTANCE MARKERS IN THE PROGENY OF NEUROSPORA CRASSA GENETIC CROSSES

After cultures from the progeny ascospores of a genetic cross containing nutritional markers are obtained, it is often necessary to test them on a variety of media. Testing using individual culture tubes may be ideal, but is time consuming. In contrast, inoculation of plates with conidial suspensions is relatively quick.

Tatum originally reported that medium containing tergitol results in an initial phase of colonial growth (Tatum et al. 1949. Science 109:509-511), and Springer reported the use of tergitol as a substitute for sorbose in N. crassa transformation procedures (Springer, Fungal Genet. Newsl. 1991 38:92). The limitations of sorbose are that it adds expense and that it takes longer to score phenotypes on sorbose-containing media due to reductions in cell growth rate. In crosses testing the segregation of arg-12s, pyr-3, and an arg-2-hph fusion gene, we found that the sensitivity of arg-12s pyr-3 progeny to arginine in the absence of uridine, the uridine requirement of pyr-3 progeny, and the hygromycin-resistant phenotype of progeny containing an arg-2-hph fusion gene could be examined using plates containing tergitol. Tergitol containing media may be useful to other investigators for large scale nutritional screening.

Our procedure was as follows: sucrose-containing plating media (containing Vogel's Medium N (Vogel, 1956. Microbiol. Genetics. Bull. 13:42-43)) and appropriate supplements were prepared and autoclaved in 100 ml aliquots in milk dilution bottles for 20 minutes. After cooling media to 50oC following autoclaving, Tergitol NP-10 (Sigma) was added to a final concentration of 0.005% by volume. Tergitol was not sterilized prior to use. Tergitol clings to the inside of the pipet tip and flushing of the tip with the sterile medium was essential in dispersing the entire volume of tergitol. Bottles of media were then gently mixed to avoid foaming and poured into standard 100 mm diameter plastic petri dishes at approximately 20 ml of medium per plate. Plates were dried prior to inoculation.

Grids were printed on acetate transparency film and taped to the bottom of the petri dishes. The grids consisted of four rows of five columns (20 squares); individual squares in the grid were approximately 13 mm x 13 mm.

Conidia were collected with a wet inoculating loop, transferred to 1 ml of sterile water, and briefly mixed with a vortex mixer. Six microliters of each suspension were applied as a spot to a square in the same grid position on each of the media involved in the screening process. Once all positions were inoculated and all liquid absorbed by the media, plates were incubated at 32oC. Growth was checked after 12, 15, 18, 24 and 48 hours of incubation. Scoring of a few progeny could be done after 15 hours, but reliable screening required 24 hours. Growth remained in the form of colonies even on extended incubation, but the aerial spreading along the lid of the petri dish which occurred was seen as potentially compromising the integrity of data collected after 48 hours.

LINKAGE AMONG MELANIN BIOSYNTHETIC MUTATIONS IN COCHLIOBOLUS HETEROSTROPHUS

Melanin is synthesized by C. heterostrophus from acetate via pentaketide and several dihydroxynaphthalene intermediates (Tanaka et al. 1991 Mycol. Res. 95:49-56), as it is for certain other fungi (Bell and Wheeler 1986 Ann. Rev. Phytopathol. 24:411-451; Kubo et al. 1989 Exp. Mycol 13:77-84; Chumley and Valent 1990 Mol. Plant-Microbe Int. 3:135-143). Previously, five melanin deficient mutants of C. heterostrophus were analyzed by Tanaka et al. (Mycol. Res. 95:49-56), who were unable to establish complete linkage relationships because three of the mutations (alb1, alb3, and brn1) showed no recombination when crossed to each other, and were unlinked to the other two (sal1 and pgr1), which mapped about 12 cM apart.

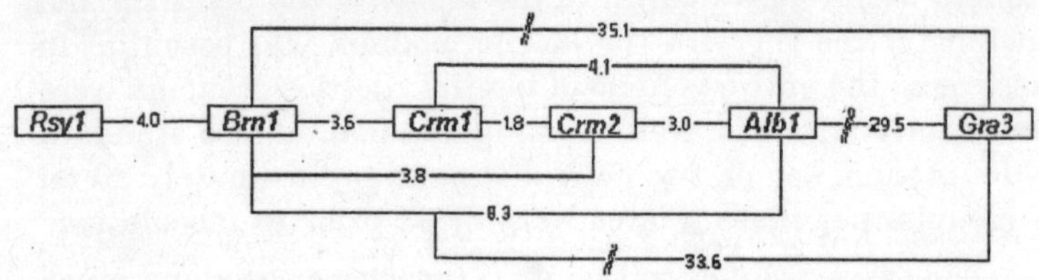

Figure 1. Proposed map of the gene cluster encoding melanin biosynthetic enzymes on chromosome 1 of C. heterostrophus, based on the two point cross data in Table 1. The gene order and the map distances must be considered tentative, since two point data may not accurately reflect recombination frequencies and do not prove gene order. Note that Rsy1 could be on either side of Brn1 and that linkage of Gra3 to the cluster is tenuous because of the distance between them. Numbers indicate cM.

A sixth color mutation, scr1, represented a third linkage group, but there was no evidence of its involvement in melanin biosynthesis. Independently, we have recovered six melanin-deficient mutants, one of which (alb1, Leach et al. 1982 J. Gen. Microbiol. 128:1719-1729) was included in the study of Tanaka et al. and maps to chromosome 1 on the C. heterostrophus RFLP map (Tzeng et al. 1992 Genetics 130:81-96). We report here that our remaining five melanin-deficient mutants [crm1 (light cream), crm2 (dark cream), brn1 (brown), rsy1 (rose), and probably gra3 (gray)] are linked to, but are not allelic with, alb1 (white) and constitute a gene cluster on chromosome. Since alb1 is the only mutant common to both our analysis and that of Tanaka et al., we do not know which (if any) of the remaining mutants in our collection correspond to those in the previous study.

In addition to the randomly isolated ascospores reported in Table 1, a few complete tetrads (four sets of twins/ascus) were recovered from each cross; they were predominantly parental ditypes with at least one tetratype ascus from each pair of parents (nonparental ditypes were not found in any progeny). Color patterns in tetratype asci indicated the following epistatic relationships [evidence for epistasis of brn1 over rsy1 is based on the relative high abundance of brn1 progeny from the brn1 X rsy1 cross, not on tetratype asci]:

alb1 > crm1 > crm2 > brn1 > rsy1 > gra3

An attempt was made to test the gene order shown in Figure 1 by scoring progeny from the three point cross alb1;brn1 X crm2. Of 945 progeny isolated, 478 were alb1, 431 were crm2, 24 were brn1, and 12 were wild type. These data are not consistent with the gene order Brn1-Alb1-Crm2, which would require that the Brn phenotype represent double cross overs as the least frequent class, which it is not. However, because epistasis obscured certain recombinant classes of progeny, the data do not distinguish between the other two possible gene orders: Brn1-Crm2-Alb1 and Crm2-Brn1-Alb1.

Although albino (alb1) strains of C. heterostrophus cannot survive in the field, they cause lesions qualitatively similar to those of wild type on the host plant (corn) in a growth chamber or greenhouse (Fry et al. 1984 Phytopathology 74:175-178). To determine if there are quantitative differences in virulence among color mutants or between color mutants and wild type, lesions on corn plants (in a greenhouse) caused by strains carrying alb1, crm1, crm2, brn1, or rsy1 were counted and measured, and compared with the sizes and numbers of lesions caused by wild type. No statistically significant differences were detected in any pair-wise comparison, suggesting that melanin plays no significant role in the virulence of C. heterostrophus to its host, in contrast to certain other fungi such as Colletotrichum lagenarium (Kubo et al. 1991 Mol.

Plant-Microbe Int. 4:440-445) and Magnaporthe grisea (Chumley and Valent 1990 Mol. Plant-Microbe Int. 3:135-143), which require melanin for penetration of the host epidermis.

MEIOTIC MAPPING OF RADIATION-INDUCED DNA REPAIR MUTATIONS AND ABERRATIONS IN *ASPERGILLUS NIDULANS*: SPECIAL FEATURES AND USE OF TRANSLOCATIONS

We recently described epistatic grouping of several new DNA repair genes and reported their location in the genetic map of Aspergillus. Supporting mapping data are presented here which showed the following special features; unusual patterns of recombination which identified three types of radiation-induced aberrations; useful translocation breaks some of which link up fragments of the meiotic map and will provided new markers for a physical map of the long arm of chromosome VII; poorly-fertile mutants which in heterozygous crosses produced "twin" cleistothecia, partly selfed for the other normally self-sterile parent; aberration-free strains which produced consistent recombination frequencies without chiasma interference, possibly because they were practically isogenic.

The origin and isolation of the mapped mutations have been reported years ago, *musK* - *musS* by Käfer and Mayor (1986 Mutat. Res. 161:119-134) and *uvsI* by Han et al. [1983 Korean J. Env. Mut. Carcin. 3:21-33]. To obtain null mutations caused by chromosome breaks, the mutagen-sensitive mutants had been induced by high doses of radiation (5-15% survival). About two aberrations per survivor had been identified in the original *mus* strains, but it is now clear that only 7 aberrations were induced by radiation in the 10 strains analyzed. A spontaneous I;III translocation was found in the strain used for isolation of mutants (FGSC #A605) as confirmed in pulsed field gels (Brody et al. 1991 Nucl. Acids Res. 19:3105-3109). The translocation breaks of this spontaneous T2(I;III) were mapped on the left arms of chromsomes I and III (Fig. 1) which complicated the mapping of mutations on these chromosomes. Strains of the 10 mapped DNA repair mutations with useful markers are now available, 6 of them translocation-free and 4 associated with an aberration (FGSC A828, A838, and A840-847).

For general mapping of mus mutations and aberrations, mitotic analysis of heterozygous diploids was combined with testing of random ascospore progeny, including aneuploids and duplications from heterozygous translocation crosses (Käfer 1977 Adv. Genet. 19:33-131). In brief, suitable segregants from the original mapping diploids were crossed repeatedly to standard strains with potentially linked markers (translocation-free segregants of *musK*, *N*, and *uvsI* for 2-4 generations, and all others for 8 or more). Recombinants were checked for translocations by testing haploids from heterozygous diploids for inter-chromosomal linkages or by monitoring crosses for increased aneuploid frequencies. In this way either translocation-free progeny were obtained or associated aberrations identified. For the latter cases, chromosome breaks were located to chromosome segments using homozygous diploids and mutations were mapped by meiotic linkage to markers on the two chromosomes involved.

Results for musM suggested that this MMS-sensitive mutation was associated with an intra-chromosomal aberration. It clearly was located on chromosome VI but showed meiotic linkage with all markers of this linkage group (LG) VI even with 3 unlinked ones (*bwA*, *sB* and *sbA*). Associated translocations were identified in 3 other cases, *musP*, *O*, and *S*. The respective mutations showed meiotic linkage to markers of two linkage groups and it is not known in which of the two chromosomes the gene is located. In the case of *musP234*, the translocation was found to be unidirectional [T2(IIL<-VIIR)]. No duplication segregants were seen among haploids from *musP*/+ diploids, but in low-density platings of ascospores about 30% of colonies were probably duplications.

They produced sectors which segregated for markers of the distal half of VIIR (*choA* and *nicB*). This finding and the map location of the translocation breaks (distal on IIL and in the middle of VIIR) are consistent with a unidirectional T2(IIL<-VIIR). Two other cases were reciprocal translocations between chromosomes III and VII. The breaks of *musO226* T2(III;VII) mapped fairly distal, especially on VIIR where *musO* was located in the most distal gap of the meiotic map of LG VI. In contrast, *musS224* T3(III;VII) mapped closer to the centromere, especially on IIIR, as identified in homozygous diploids.

In these final aberration-free crosses, linkage values were very consistent and agreed well with previously published ones. Average recombination frequencies between all possible pairs of markers are presented for 4 cases, *musK, R, Q,* and *L* and in 2 of them raw data for one cross are shown as well. Mutations could be located unambiguously in most cases; e.g., *musK228* was mapped distal to all markers in LG VIIIR, based on practically identical results from 2 crosses which showed good additivity. Only the location of *musR223,* very close distal or proximal of *dilA*, is not yet known exactly, even though standard errors were small and linked aberrations had been eliminated; because only fairly distant other markers segregated in the cross to *dilA*, single and double crossovers were equally frequent. Similarly, results were ambiguous in the first mapping cross of *musQ* in which results suggested a position distal to *adC* (evident in cross 2284).

However, crosses with the additional marker *acrB* clearly placed it proximal to adC. Results for *musL* were complicated in early crosses by linkage to the spontaneous translocation T2(I;III), and in general by poor viability and recovery of *musL*. Among ascospores, the frequency of *musL* was about 15%, but in recorded results allele ratios were usually higher, because the poorly-growing musL progeny tended to be isolated preferentially. In addition, fertility of *musL* was very poor in heterozygous crosses [e.g., crosses to more distal markers were not successful]. This presumably caused the observed unusual production of "twin-type" cleistothecia; i.e., a mixture of 2 sets of ascospores, one of them hybrid and segregating for *musL*, the other consisting of selfed progeny of the second parental strain. These problems made accurate location of *musL* distal to *yA* and *biA* impossible.

When the patterns of double or triple crossing over were analyzed, almost complete absence of chiasma interference was evident in all crosses; e.g., one class of the single crossover (CO) types was often less or equally frequent as the corresponding double CO types. This suggests random distribution of chiasmata without interference, as evident in the crosses of uvsI. The data from *musN* crosses with more distant markers are less conclusive because in such crosses interference is diminished in all species (Foss et al. 1993 Genetics 133:681-891).

In summary, meiotic mapping revealed that in 4 cases radiation-induced aberrations were associated with mus mutations which presumably are null-mutations. Three types of aberrations were identified: an inversion, a unidirectional translocation, and two reciprocal translocations. Such aberrations reduce meiotic recombination frequencies in heterozygous crosses and are exploited for genetic mapping in various species. The postulated inversion *musM225* (AbVI-2) is not yet mapped well enough to be used in this way. It resembles the known inversion, *lysA1* (AbVI-1) located in the same LG, which can be detected by high aneuploid frequencies in crosses to overlapping translocations.

On the other hand, three translocations that have been mapped will be useful. The unidirectional translocation of *musP234* transfers the distal half of the right arm of chromosome VII to the tip of IIL. When crossed to mutations assigned to LG VII, these will be disomic in the resulting duplications only if they are located distal to the break point (near *malA*). However, it will be important that progeny from such crosses not be used for standard analysis, since in this case duplications and their sectors are not verydistinct and near-haploid products from duplications may accumulate further aberrations (Roper and Nga 1969 Genet. Res. 14:127-136). Two reciprocal translocations both involve chromosomes III and VII.

One of them, *musS223* T3(III;VII), will be a useful centromere marker on IIIR, while the other, *musO226* T2(III;VII), provides meiotic linkage between *choA* and *nicB*, connecting the most distal segments of VIIR. This should make it possible to orient the markers of the *nicB* fragment (as recently achieved for the distal IIR region in fission yeast; Egel 1993 Curr. Genet. 24:179-180). Several previous translocations were mapped in LG VII and these were essential for the mapping of the centromere and the ordering of 7 meiotically unlinked fragments (Käfer 1977, ref. cit.). Even now the standard meiotic map of linkage group VII has 6 gaps with 40-50 % recombination and translocation breaks have so far been placed into only 3 such gaps; therefore 3 gaps without linkage remain in the meiotic map of VIIR.

The location of all *mus* mutations which were separable from aberrations could be identified accurately, except for *musL*.

Recombination between *musL* and *yA* or *biA* varied widely among progeny of different color, because samples were non-random for *musL* as well as *yA*. In addition, mapping was inaccurate because "twin" cleistothecia were frequent (as found also for other poorly fertile mutations; e.g., *bimD*; Denison et al. 1993 Genetics 134:1085-1096). Such twins were "dizygotic" and contained selfed ascospores of the other parent, even if this strain was normally self-sterile (e.g., *ribo-*). It seems likely that in these crosses fertility is increased by cross-complementation. The partial selfing further increased the uncertainty of linkage values even when several unlinked markers segregated and purely parental types, could be identified.

Aberrations or mutations which reduce recombination and reveal meiotic linkages are especially helpful in species with genetically large chromosomes; i.e., not only for our standard strains which are all derived from backcrossed strains that originated from a single nucleus (Pontecorvo et al. 1953 Adv. Genet. 5:141-232) but also for strains of fission yeast which had a similar origin. In this yeast, as in *A. nidulans*, recombination values are large and practically without interference and long-range mapping requires special recombination-defective strains (Schmidt 1993 Curr. Genet. 24:271-273). In both species strains are therefore isogenic barring induced mutations. In contrast, species which show chiasma interference, such as Neurospora or Drosophila, depend on crosses of less closely related strains of opposite mating type or sex. High recombination without interference may therefore be related to isogenicity, especially since highly backcrossed strains of *N. crassa* consistently showed higher recombination and less interference (Käfer 1982 Neurospora Newsl. 29:41-44). Evidnce for or against this hypothesis is starting to accumulate as techniques are becoming available for measurements of divergence at the nucleotide level and in some cases effects of heterology on recombination and on related processes have been demonstrated.

3

Lethality or Improved Growth

Genetic analysis recently identified nine new DNA repair genes of *Aspergillus nidulans*, *musK* - *musS*, which map on several different chromosomes (Käfer 1994, preceding article in this Newsletter). Such *mus* mutants are sensitive to certain chemical mutagens, but not sensitive or only slightly sensitive to UV and gamma-radiation. To identify epistatic interactions with members of the 4 Uvs groups, double mutants strains were isolated (Käfer and Chae 1994 Curr. Genet. 25:223-232). However, some *mus;uvs* double mutants could not be analyzed because they grew too poorly or were lethal (showing "synthetic enhancement in gene interaction"; Guarente 1993 Trends Genet. 9:362-366). The opposite effect was also found; namely interaction which led to improved recovery and growth (or "rescue") as documented here.

Double *mus;uvs* mutants were isolated as haploid segregants from heterozygous diploids when genes mapped on different chromosomes, or by random ascospore analysis of intercrosses in cases of linkage. Heterozygous diploids usually carried markers linked in repulsion to *mus* and *uvs* mutations. Samples of haploid segregants (50 to >300) were tested for nutritional markers and for growth on MMS, to which all mutants except *uvsI* are sensitive. In most cases, these tests identified the 3 mutant types, including the double mutant segregants which showed increased mutagen sensitivities in several cases (25% each).

In some cases recovery was poor even for single mutants (especially for *musL* or *uvsB* segregants). To obtain sufficiently large samples of poorly viable segregants, linked color markers or associated phenotypes were used to "select" the missing types (e.g., segregants were isolated preferentially for *musL, musQ, musK* and *uvsC,* which are linked to *yA, wA,* or to *fwA* and *chaA,* respectively; similarly, *uvsB* and *D,* or *musO,* could be recognized because they show dark mycelial color, or poor conidiation). In such selected samples, only single and double mutant segregants are expected (50% each).

When marker segregation suggested that double-mutant segregants were much rarer than expected, additional haploids from at least two diploids were analyzed (one case with an aberration as the cause of low recovery was identified in this way; *musQ;uvsH.* In some cases, very rare segregants with markers expected of double mutants were crossovers, especially when recovery of single mutants was normal. Such exceptions, when checked on benomyl, occasionally turned out to be diploid (e.g., *musO/+;uvsH/H,* which was useful for the isolation of *musO;uvsH* double mutants). In all cases, rare presumptive doubles and cases with ambiguous phenotypes were tested by outcrossing or by complementation to confirm the double mutant genotype.

For some pairs of *mus* and *uvs* mutations which showed meiotic linkage, double mutant crossover types were obtained by selecting against linked markers used in repulsion (especially in crosses of the III;VII translocations associated with *musO* or S to *uvsI* which maps on chromosome III). Double *mus;uvsI* mutants from such crosses were identified as MMS-sensitive progeny which also showed the high UV sensitivity typical for uvsI (Chae and Käfer 1993 Curr. Genet. 24:67-74).

All types of isolated segregants are listed in Tables 1-3, and the recovery of double mutants is shown in percent relative to the "less viable" single mutant type. In general, only double mutants with good growth showed close to 100% recovery (e.g. *mus;uvsI* doubles), because of the competition between different segregant types on the benomyl plates during haploidization. "Lethal" interactions are assumed when practically no double mutants but significant numbers of both single mutant parents were obtained (e.g., for *uvsF* when combined with *musN,* or in for *uvsF;musQ* doubles). In random samples, certain *musL* and *musO* double mutants also could not be recovered, mainly because single mutants showed very poor recovery. However, when single mutants were selected, a significant fraction of doubles were obtained among them.

Clearly the majority of lethal interactions were found for mus mutations interacting with *uvsF* (5 of the 6 observed cases); only one other case is completely lethal (*musO* with *uvsC*). Unexpectedly, none of the *mus;uvsH77* double mutants was completely lethal and only two showed very poor recovery, even though *uvsH* is a member of the UvsF epistatic group. In contrast, the two members of the UvsB group showed very similar results. Both *uvsB* and *D* affected viability of double mutants in all cases, but in two opposite ways for different *mus* mutations. Double mutants either were growing more poorly than *uvsB* (or *D*) singles and were produced with very low frequencies; or they showed better growth than *uvsB* (or *D*) and much increased recovery (>200 - 400 % for *musN, P* and *R* double mutants).

Lethal interactions between *uvs* mutations had been found previously appeared to be the rule for mutations of different epistatic groups in Aspergillus (Käfer and Mayor 1986 Mutat. Res.161:119-134). However, viable double mutants of *uvsI* with other *uvs* mutations have been identified recently, even for intergroup pairs (Chae and Käfer 1993 Curr. Genet 24:67-74) as also found here for the many cases of *mus;uvsI* doubles. The latter results resemble the findings for double *rad* mutants of budding yeast, which usually are viable in all combinations. Only a few "synthetic lethals" have been found for radiation-sensitive mutants of this species; e.g., for certain *rad52* mutations, when combined with unusual alleles of the excision repair gene *RAD3*, or of the topoisomerase gene *TOP1* (Montelone et al. 1988 Genetics 119:289-301; Levin et al. 1993 Genetics 133:799-814). However, for many other important mutants, interacting mutations which lead to enhanced effects and lethality have recently been reported, and selective systems have even been devised (Bender and Pringle 1991 Mol. Cell. Biol. 11:1295-1305).

An interesting finding of the opposite type is the interaction of certain *mus* mutations with UvsB group mutations; namely, in double mutants with *musN, P* and *R*, suppression of the low growth rate and poor viability typical for UvsB group mutants was found. Such double mutants were also recovered more frequently (3-8x) and showed lower MMS sensitivity than single *uvsB* (or *D*) segregants (but only at low concentrations; Käfer and Chae 1994, ref. cit.). Furthermore, the same effect was observed in triple mutant strains. Growth rate and recovery improved (2-4 x) when *musN* was incorporated into double *uvs* strains containing a UvsB group mutation (e.g., from one such diploid, 18 *musN;uvsC;uvsD,* plus 66 *musN;uvsD* segregants were recovered, compared to 1 *uvsC;D* plus 19 *uvsD* not containing *musN*). In addition, triple mutant *musN;uvsC;uvsD* strains conidiate and grow much better than *uvsC;uvsD* double mutants which have extremely poor growth and conidiation. These findings are analogous to those known for certain *recA* mutants of *E. coli* in which the recBCD enzyme recklessly degrades DNA, while *recA;recB* double mutants do not show such effects.

SENESCENCE IN STRAINS OF NEUROSPORA FROM SOUTHERN INDIA

Natural isolates of Neurospora which cannot be vegetatively propagated continuously have been identified among collections from distant geographical locations (X. Yang and A.J.F. Griffiths 1993 Mol. Gen. Genet. 237:177-186). We have found that senescence is common in strains of N. intermedia growing on burned sugar cane in fields in Maddur, approximately 80 km southwest of Bangalore. In a survey of 150 strains, 29 strains have died to date in 57 serial subcultures made at intervals of 5-10 days on slants of Vogel's N medium + 1.5% sucrose. Senescence was first observed after a minimum of 9 subcultures. Nearly 30 more strains have lost their vigor or have become aconidial.

The expression of senescence in most strains occurred abruptly. In a subculture series, conidial transfer from a previously healthy looking culture resulted in only a thin mycelial growth which failed to produce aerial hyphae and conidia. The surface mycelium ceased growth altogether in 1-3 subsequent transfers and considered dead.

To determine if the time of death is reproducible, a duplicate subculture series was made with some identified senescence-prone strains. These strains had been stored at - 20oC for 12 months as conidial cultures. As shown in Table below, there was considerable variability among the strains with regard to the subculture number in which senescence was expressed. In all cases, senescence in the second series occurred after fewer passages. In one instance, an identified senescence-prone strain did not resume growth from the frozen stock culture. These observations indicate the possibility of molecular/subcellular degenerative changes occurring even under conditions of apparent dormancy.

Table. Subculture number in which death occurred in some strains of N. intermedia

Strain	Series 1	Series 2
Maddur 1991-1	21	16
Maddur 1991-3	25	17
Maddur 1991-59	29	14
Maddur 1991-60	15	12
Maddur 1992-18	29	9

Inheritance of the senescence phenotype and the time of its expression was studied in the progeny of a successful cross between two senescent strains when crossed after two different numbers of passages. In the initial series, strains Maddur 1991-21a and Maddur 1991-3A had become senescent in the twentieth and twenty-fifth subcultures, respectively. These parents were crossed prior to terminal subculture by inoculating them together on a synthetic crossing medium containing filter paper as the carbon source. Forty random ascospore progeny were analyzed of which 39/40 (98%) died in 3-9 subcultures; the average subculture number in which death occurred was five. When the cross was repeated with stocks which had been subcultured only three times before, only 3 out of 40 progeny cultures died in up to the tenth subculture. These observations are similar to those of A.J.F. Griffiths and H. Bertrand (1984 Curr. Genet. 8:387-398) and are indicative of progressive damage with subculturing. Yet, surprisingly, when the parents were crossed prior to the terminal stage, the sexual progeny was invariably rejuvenated at first. This observation, and the fact that our senescent strains came from the same locality where we have found sexual reproduction to be prevalent, suggests that senescence in nature is maintained through outbreeding.

ALTERNATE WAYS TO PRESERVE STRAINS WITH SILICA GEL

The method for storing Neurospora strains described by D. D. Perkins (Can. J. Microbiol 8:591-594, 1962; Neurospora Newsl. 24:16-17, 1977) and elaborated in a collection of articles in Neurospora Newsl. 26 has made it possible to keep large collections with much less effort than would be required with the older lyophil method. In Perkins' method, conidia are suspended in sterilized non-fat milk, and the suspension is pipetted onto chilled sterile silica gel. For non-conidiating strains, mycelia are mulled or otherwise fragmented in milk to make the suspension. However, even this greatly improved method requires a non-trivial amount of manipulation when large numbers of strains are to be preserved, especially for non-conidiating strains. Preparing each stock consumes a pipet and at least two test tubes: one for growth of the strain, and one for preservation. Both tubes need to be labelled, which adds to the effort and to the chance of error. The following method requires no pipetting. The stock is preserved in the tube in which it was grown.

Perlite, a fluffy white volcanic ash available at any garden center, is put into culture tubes up to no more than 1/5 of their total length. A liquid medium of choice (see below) is added so as just to cover or nearly cover the Perlite and the tubes are autoclaved. The tubes are inoculated and then tapped in a near-horizontal position to form slants which holds their form when the tubes are gently returned to vertical. When the strains have conidiated well, or if non-conidiating, have grown to their limit, baked silica gel, 6-12 mesh, is added to about 4/5 the capacity of the tubes. The tubes are capped, sealed with pre-cut strips of Parafilm, and put at 4°C. At least two weeks later, they are shaken vigorously and stored at 4°C or in a deepfreeze. (The addition of silica gel should be done in a hood for obvious reasons. I have found it convenient to sterilize and dry the gel in an oven in a casserole or open beaker for two hours at 450°F (about 230°C) and store it in pre-sterilized bottles. Sterile 3 ounce paper dixie cups, upside down and separated by squares of toilet paper, are convenient single-use vessels, and the cups are squeezed to form a pouring lip for adding the gel to the tubes.) A fragment of the dry Perlite or a crystal or two of silica gel will start a new culture of a conidiating strain. For non-conidiating strains, the Perlite is necessary; silica gel crystals usually remain sterile. If the Perlite is difficult to disperse, breaking it up the mass with a sterile bamboo stick can be helpful.

Recently I have found that "seed beads" are a very convenient alternative to Perlite, especially for non-conidiating cultures. I use very small black or colorless beads from the Czech Republic, available at craft shops or at Discount Beads, POB 186, The Plains OH 45780; tel. 1-800-793-7592; $25/kilo. I have used 5 g of beads per 18x150 mm tube with 1.5 ml of medium, and form slants by tapping, as with the Perlite. (Presumably, 13x100 mm tubes with about 0.75 ml of medium would work as well.) The holes in the beads trap medium and the mycelium grows into these protected holes. The beads disperse more easily after they are dried with silica gel than does Perlite. The convenience of seed beads often seems worth the greater cost, still only 12.5 cents per tube.

Finally a caveat. I have not had trouble with cultures dying, but my experience has been very short: about a year with most of the Perlite cultures, and negligible with the seed-bead cultures. Only time will tell how the longevity of these stocks compares with that of stocks made in the traditional ways. Others have found that non-fat milk as a suspending medium enhances survival of lyophil cultures. I have found that Neurospora grows well in otherwise permissive liquid medium to which non-fat powdered milk has been added to 10% w/v before autoclaving, and I have used this with beads to prepare stocks. However, I do not know whether this improves survival of the stocks. I would appreciate hearing the experience of anyone who decides to try this method.

PHENOTYPIC REVERSION OF ASPERGILLUS NIDULANS MORPHOLOGICAL DETERIORATED VARIANTS IN THE PRESENCE OF OSMOTIC STABILISERS

Strains of Aspergillus nidulans with chromosome duplications are unstable at mitosis. They produce sectors which are mainly of two types: improved sectors that result from partial or total loss of the duplication segment and deteriorated sectors having poor conidiation and dark brown mycelium. It is postulated that deteriorated variants carry additional duplications resulting from non-homologous sister-chromatid exchange within the duplicated segments. (Nga and Roper, 1968 Genetics 58: 193-209). Deteriorated sectors are unstable but can give more derivatives which probably are the result of transpositions of the tandem duplication segment to other regions of the genome (Azevedo and Roper, 1970 Genetical Research 16: 79-93). Crosses between these more stable deteriorated variants are not always successful due probably to incompatibility factors. When protoplast fusion was attempted to cross some incompatible deteriorated variants, protoplasts were regenerated on medium with 0.6M KCl as osmotic stabilizer.

All colonies on this medium had a normal phenotype but, when transferred to medium without the osmotic stabilizer, the deteriorated phenotype returned. It was then supposed that in the presence of the osmotic stabilizer deteriorated variants are phenocopies of a normal strain. Other osmotic stabilizers were tested (1M sucrose; 0.5M MgSO4; 1.2M sorbitol) and in all cases normal or almost normal phenotypes were observed.

The original duplication strain that gives rise to deteriorated variants (strain A), a non diplication strain arising as a sector from strain A, several deteriorated variants and one aneuploid strain, were tested in relation to the number of conidia produced on medium with and without 0.6M KCl. Table I shows that all strains produced more conidia on medium with 0.6M KCl. However, the increased conidiation observed with deteriorated variants was greater and resulted in colonies with normal or almost normal phenotypes.

Growth rate was not significantly altered on medium with stabilizer when compared to the growth rate on medium without osmotic stabilizers. Media containing osmotic stabilizers are being used in our laboratory to distinguish deteriorated sectors emerging from duplication strains, from other types or sectors with altered morphology, mainly aneuploids. Deteriorated sectors with characteristic morphology - brown mycelium and poor conidiation - typically revert to normal morphology on high osmolarity medium. Other sectors such as color variants and aneuploids do not show reversion. In meiotic crosses, and mitotic analysis, osmotic stabilizers are also being added to the medium to make easier conidial color distinction of deteriorated segregants.

NEW PLASMID AND LAMBDA/PLASMID HYBRID VECTORS AND A NEUROSPORA CRASSA GENOMIC LIBRARY CONTAINING THE BAR SELECTABLE MARKER AND THE CRE/LOX SITE-SPECIFIC RECOMBINATION SYSTEM FOR USE IN FILAMENTOUS FUNGI

In the previous Fungal Genetics chapter, we described a series of plasmid vectors constructed carrying the bar gene as a selectable marker for use in filamentous fungi (Pall and Brunelli 1993 Fungal Genetics Newsl. 40:59-63; Pall 1993 Fungal Genetics Newsl. 40:58). In this note, we describe an additional plasmid expression vector carrying this selectable marker and the construction of four llambda/plasmid hybrid vectors carrying the bar gene within plasmid inserts that can excise by Cre/lox-mediated excision. A Neurospora crassa genomic library constructed in one of these lambda/plasmid hybrid vectors is also described below.

We have inserted the grg-1, glucose-repressible promoter (McNally and Free 1989 Curr. Genet. 14:545-551) as a 1.4 kb NcoI to BamHI fragment from pDF255, into a bar- containing expression vector in order to provide a regulatable promoter. The vector produced was designated pBARGRG1 (Fig. 1). The grg-1 gene has recently been shown to be identical to the ccg-1 clock controlled gene (S. Free, personal communication; Loros and Dunlap 1991 Mol. Cell. Biol. 11:385-388) so this promoter is also regulated by the circadian rhythm clock. It should be noted that this promoter has been patented (State University of New York) and is licenced to the Hawaii Biotechnology Group so permission must be obtained in order to use it for commercial purposes. As in the case of our other expression vectors, this one is designed to express coding sequences under their own ATGs. The pBARGRG1 plasmid is larger than those described earlier (Pall and Brunelli, op. cit.) because a larger promoter fragment is necessary in order to maintain the glucose repression (S. Free, personal communication).

This vector and the pBARGEM7-2, pBARMTE1 and pBARGPE1 vectors all contain a lox-NotI-lox fragment, as described earlier, allowing these to be attached to lambda arms, generating lambda/plasmid hybrid vectors in which the plasmid sequences can be efficiently excised by Cre/lox-mediated site-specific recombination (Brunelli and Pall 1993 Yeast 9:1309-1318; Pall and Brunelli, op. cit.). The lambda arms used for this are the same as those used previously (Brunelli and Pall, op. cit.) but a novel method of plasmid insertion was used that will be described elsewhere (Brunelli and Pall, BioTechniques, in press). All four of these lambda vectors have been constructed, and are designated as shown in Table below.

Table. Plasmids inserted into lambda/plasmid hybrid vectors

Plasmid insert	designation
pBARGEM7-2	BARGEM7
pBARMTE1	BARMTE1
pBARGPE1	BARGPE1
pBARGRG1	BARGRG1

These vectors when infected into an E. coli lysogenic for lambda KC (Elledge et al. 1991 Proc Natl. Acad. Sci. USA 88:1731-1735), carrying the cre gene and a kanamycin resistance marker, excise the plasmid inserts by Cre-lox-mediated recombination (automatic subcloning). The plasmid that is excised differs from the initial plasmid in that it only contains one lox site. As suggested earlier, we designate single lox plasmids with a prime to distinguish them from the double lox plasmids, i.e., pBARGRG1'. Another lambda/plasmid hybrid vector designed for use in Aspergillus nidulans was described recently by Holt and May (Gene 133:95-97, 1993).

It should be noted that the EcoRI and XhoI sites of the polylinker sequences of these plasmids and of the lambda vectors containing them are unique, allowing them to be used for construction of lambda genomic or cDNA libraries. A Neurospora crassa genomic library was constructed using DNA from wild type strain 74OR-23-1A as follows:

Neurospora DNA was subjected to partial digestion using the restriction enzyme Tsp509I (New England Biolabs), an enzyme that restricts 5' to the sequence AATT. This produces compatible single stranded ends to that produced by EcoRI cut

DNA. Lambda DNA isolated from BARGEM7 was ligated to form concatemers, cut with EcoRI, and phosphatased using calf intestinal phosphatase. The partially restricted Neurospora DNA was run on a low melting point agarose gel by electrophoresis and the DNA was separated by size and isolated using gelase. The Neurospora DNA was then ligated to the lambda DNA and packaged using commercial packaging mix (Stratagene).

A library containing approximately 2.5 X 10^6 plaques was produced with a background of 3% blue plaques (not containing inserts). When individual members of the library were excised as plasmids and restricted to determine the size of the inserts, inserts ranging from 2 kb to over 10 kb were found, averaging about 5 to 6 kb. It should be noted that this vector can package inserts of up to 11 kb in size, so these inserts are averaging about half the maximum possible size. With an estimated average insert size of 5.5 kb and Neurospora crassa genome size estimated at about 40 megabases (Perkins, 1990 Fungal Genetics Newsl. 37:9), this library should contain each Neurospora DNA sequence about 340 times.

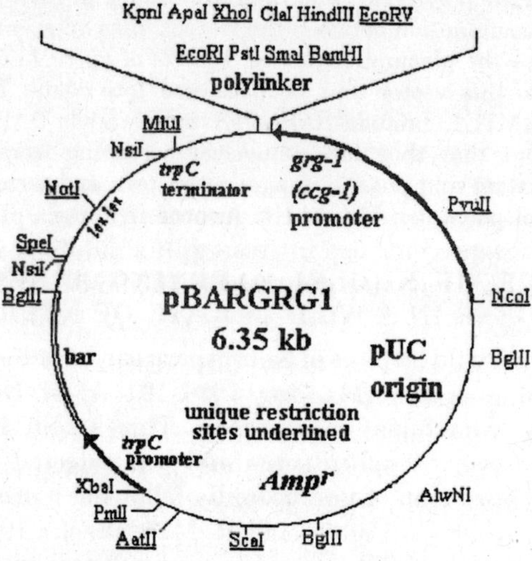

Figure. Fungal expression vector carrying the grg-1 (ccg-1) promoter. This vector also carries a sequence for Ignite/Basta-resistance (bar) that functions in fungi and in E. coli and the lox-NotI-lox sequence for site-specific recombination.

Plasmid and lambda vectors and the Neurospora genomic library are all available from the Fungal Genetics Stock Center. We thank Dr. Steve Free for supplying plasmid pDF255 containing the grg-1 promoter. This research was supported by the College of Sciences at WSU.

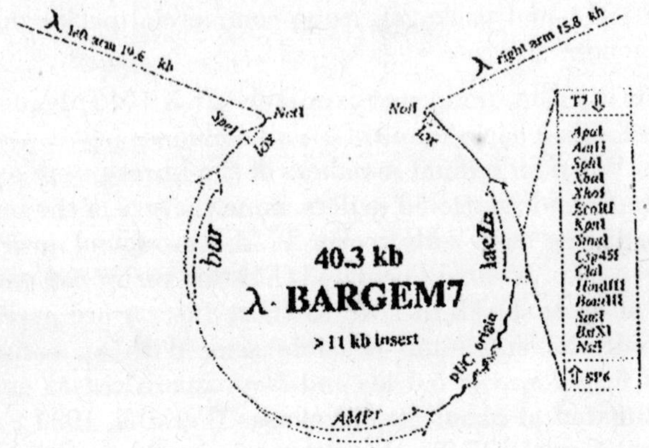

Figure. Diagram of the BARGEM7 vector. The restriction sites EcoRI and XhoI (in the polylinker) and SpeI are unique in BARGEM7. Cre-mediated recombination between the two lox sites adjacent to the two arms releases the plasmid sequence. Inserts of up to 11.5 kb can be inserted into this vector and be packaged into coats. The vectors lambda-BARMTE1, lambda-BARGPE1 and lambda-BARGRG1 are similar except that they lack blue/clear screening from lacZalpha complementation and possess fungal promoters and terminators for expression of polylinker-inserted sequences in fungi.

PRESENCE OF NUCLEI CARRYING A RECESSIVE LETHAL GENE IN A WILD ISOLATE OF NEUROSPORA

To learn if wild isolates of Neurospora can be heterokaryotic, we studied a strain (Maddur 1991-101 A) of Neurospora intermedia from burnt sugar cane. This strain has grown normally in over 50 subcultures and was selected because it produced a pure crop of microconidia following a procedure we described earlier (A. Pandit and R. Maheshwari 1993 Fungal Genet. Newsl. 40:64-65). Homokaryotic cultures were derived by plating microconidia produced by mycelium grown over cellophane agar. Of 159 homokaryotic cultures, one (Maddur 1991-101 A-10) was immediately noticeable because of its poor growth and conidiation when transferred to fresh medium. After another transfer its vigor declined further. Serial subculturing resulted in progressive decline in growth rate and in the fourth subculture it ceased growth altogether. Thus the homokaryotic derivative, Maddur 1991-101 A-10, showed a senescent phenotype. It was crossed to a wild-type N. intermedia (FGSC 3417). Progeny from reciprocal crosses were serially subcultured at weekly intervals. It was observed that the senescent phenotype in both crosses was inherited in a 1:1 ratio showing the mutant form was due to a single nuclear gene. A cross between two senescence-prone fl cultures yielded progeny which all died in 5 to 11 serial subcultures. It was noted that initially all ascospore-derived cultures conidiated profusely even though the parents has lost vigor and had poor conidiation at the time crosses were made. Moreover, the number of passages before death was manifested varied in progeny subcultures, suggesting the presence of genes which influence the expression of the mutant lethal gene. The mutant lethal gene has been successfully introgressed into N. crassa through six backcrosses and has been maintained in heterokaryons.

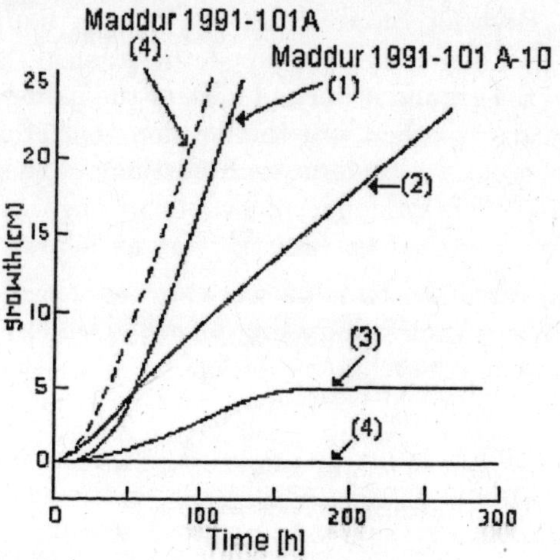

Fig. Comparative growth of the parent heterokaryon, Maddur 1991-101 A (broken line) and of the homokaryon, Maddur 1991-101 A-10 (solid line), in race tubes. The subculture number is given in parentheses.

This study shows that a recessive lethal gene has arisen and persists in wild-collected Neurospora but is masked in the heterokaryotic mycelium until it segregates in the spores.

SEXUAL REPRODUCTION BY NEUROSPORA IN NATURE

Except for a solitary report on the presence of perithecia of Neurospora on pine trees burned in the fire which followed an earthquake in Tokyo (Kitazima 1925 Ann. Phytopath. Soc. Japan 1:15-19), all other attempts to find the sexual stage in nature were unsuccessful (Shear and Dodge 1927 J. Agric. Res. 34:1019-1024; Prakash 1965 Neurospora Newsl. 7:5-6; Shaw 1990 The Mycologist 4:6-13). However, based on the available information on the population structure of Neurospora, Perkins and Turner (1988 Exp. Mycol. 12:91-131) predicted that sexual reproduction is prevalent in nature.

We present evidence for the occurrence of sexual reproduction of Neurospora in nature. Because of the preference of Neurospora for burned substrates and the regular availability of scorched sugar cane in agricultural fields, this material was selected for examinations. The place of our study was Maddur in Mandya District, approximately 80 km southwest of Bangalore. After sugar cane is reaped, the trash of the stripped leaves left on the ground is burned to clear the fields. The sugar cane stumps are scorched, but the development of new shoots (ratoon crop) from the underground portions of these stumps is not affected. The calendar of events in sugar cane fields which had been burned in January was as follows:

About a week after the fields were burned, Neurospora was visible as distinct pink conidiating pustules which were more common at the nodes than at either internodes or the cut surfaces of the scorched stumps.

The mycelium ramified in the stump and was dense under the epidermal plant tissue. Pustules of conidia emerged at the nodes through root openings and through cracks in the epidermal tissue. The juicy stumps supported profuse and prolonged conidiation. These stumps were brought to the laboratory and examined with a binocular microscope. Perithecia of Neurospora were not found in the stumps in which Neurospora was conidiating.

One to two months after the burn (February-March), the stumps had begun to shrivel. These were surrounded by a ratoon crop which had grown approximately two feet high. Pale and withered pustules of Neurospora could still be seen on the stumps. However, these stumps also did not have perithecia either on the surface or under the epidermal tissue.

By the end of April, the new crop had grown approximately three feet, obscuring the stumps. Although pre-monsoon showers had come in the last week of April, the burned- over stumps appeared dry. In these shrivelled stumps, the epidermal plant tissue had become separated from the ground tissue and had begun to fall off. Excepting old, whitish patches of Neurospora in a few stumps, the conidial pustules had mostly disappeared. When epidermal tissue was peeled off, clusters of Neurospora perithecia were found at a few places on the ground tissue in some stumps. Most perithecia were shrivelled but a few were typically flask-shaped with beaks. In some stumps perithecia protruded through the cracked epidermal tissue. Some perithecia contained asci in different stages of development. In a few stumps, perithecia had ejected ascospores, which were sticking to the under surface of the epidermal tissue or on the ground tissue. The ascospores were grooved and measured 27 x 15 m. When ascospores were isolated on Vogel's N medium + 1.5% glucose and heat-shocked, nearly 40% germinated. The resulting cultures were identified as Neurospora intermedia, based on successful crosses with tester strains.

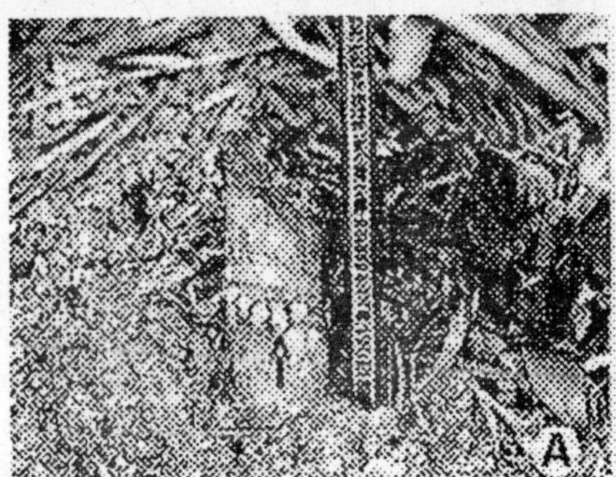

Fig. Conidiating pustules of Neurospora (arrow) emerging through root openings at the node of a scorched sugar cane stump (approximately two weeks after the burn).

We have observed perithecia in burned-over stumps of sugar cane for three successive years. The observation that asexual and sexual stages of Neurospora are temporally separated and that perithecia are found under the epidermal tissue has raised questions as to the role of conidia, the mechanism of fertilization, the mode of dispersal of ascospore and the mechanism of initiation of 'infection'. Information on these and other aspects of the life cycle of Neurospora in nature will be published elsewhere.

Fig. Shrivelled stump (arrow) surrounded by a ratoon crop approximately three months after the burn.

Fig. Light micrograph of perithecia on ground tissue partially exposed by removing epidermal tissue of the burned sugar cane stump. Perithecia (arrow) were those of Neurospora. Other perithecia were of Gelasinospora as identified by ovoid ascospores with pits under SEM. (Bar, 1 mm).

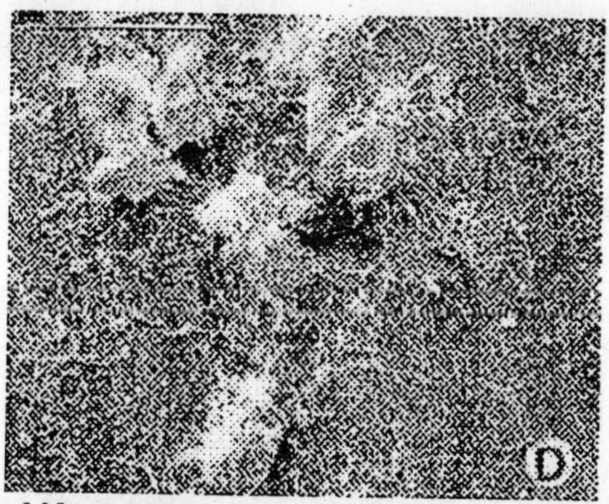

Fig. SEM of Neurospora perithecia (arrow) on ground tissue of burned sugar cane stump. (Bar, 1 mm).

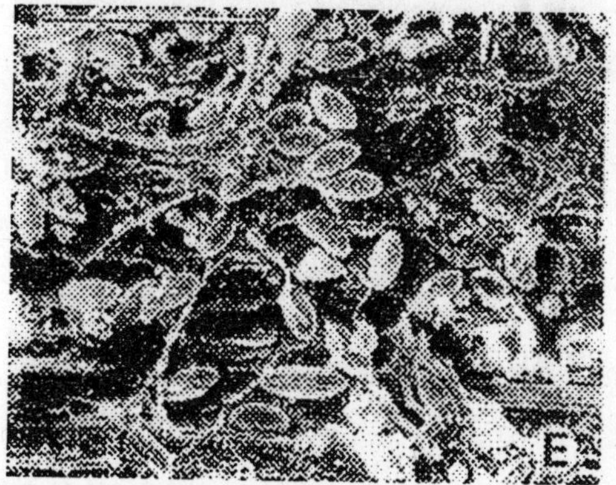

Fig. SEM of Neurospora ascospores on a stump. (Bar, 100 microns).

DEVIATIONS FROM 1:1 AND NUMBERS OF PROGENY NECESSARY FOR ESTABLISHING LINKAGE

We have found the following table useful when assigning new genes to linkage groups, either with Parental:Recombinant random isolates or with Parental ditype:Nonparental ditype tetrads. The ratios for total numbers through 50 are from Perkins, D.D. 1953 Genetics 38:187-197, and were extracted from tables in Warwick, B.L. Texas Agric. Exper. Sta. Bull. 463:1-28. The ratios for total numbers from 51-100 were obtained using binomial probability paper (Mosteller, F., and J.W. Tukey. J. Amer. Stat. Assoc. 44:174-212).

Smallest numerical ratios showing significant deviation in one direction from 1 : 1

Total	Ratios		
numbers	5%	2.5%	1%
5	5:0		
6	6:0	6:0	
7	7:0	7:0	7:0
8	7:1	8:0	8:0
9	8:1	8:1	9:0
10	9:1	9:1	10:0
11	9:2	10:1	10:1
12	10:2	10:2	11:1
13	10:3	11:2	12:1
14	11:3	12:2	12:2
15	12:3	12:3	13:2
16	12:4	13:3	14:2
17	13:4	13:4	14:3
18	13:5	14:4	15:3
19	14:5	15:4	15:4
20	15:5	15:5	16:4
21	15:6	16:5	17:4
22	16:6	17:5	17:1
23	16:7	17:6	18:5
24	17:7	18:6	19:5
25	18:7	18:7	19:1
26	18:8	19:7	20:6
27	19:8	20:7	20:7
28	19:9	20:8	21:7
29	20:9	21:8	22:7
30	20:10	21:9	22:1
31	21:10	22:9	23:8
32	22:10	22:10	23:9
33	22:11	23:10	24:9
34	23:11	24:10	25:9
35	23:12	24:11	25:10
36	24:12	25:11	26:10
37	24:13	25:12	26:11
38	25:13	26:12	27:11

39	26:13	27:12	28:11
40	26:14	27:13	28:12
41	27:14	28:13	29:12
42	27:15	28:14	29:13
43	28:15	29:14	30:13
44	28:16	29:15	31:13
45	29:16	30:15	31:14
46	30:16	31:15	32:14
47	30:17	31:16	32:15
48	31:17	32:16	33:15
49	31:18	32:17	34:15
50	32:18	33:17	34:16
51	33:18		
52	33:19		
53	34:19		
54	34:20		
55	35:20		
56	35:21		
57	36:21		
58	36:22		
59	37:22		
60	38:22		
61	39:22		
62	39:23		
63	39:24		
64	40:24		
65	40:25		
66	41:25		
67	41:26		
68	42:26		
69	43:26		
70	43:27		
71	44:27		
72	44:28		
73	45:28		
74	45:29		
75	46:29		
76	46:30		
77	47:30		
78	47:31		

79	48:31
80	49:31
81	49:32
82	50:32
83	50:33
84	51:33
85	51:34
86	52:34
87	52:35
88	53:35
89	54:35
90	54:61
91	55:61
92	55:37
93	56:37
94	56:38
95	57:38
96	57:39
97	58:39
98	58:40
99	59:40
100	59:41

Numerical ratios are those attaining the one-sided significance level shown in percent.

NEUROSPORA TETRASPERMA HELPER STRAINS USING THE E GENE

In N. crassa, strains with an inactive mating type allele are available that can be used as one component of a forced heterokaryon, serving as a helper to shelter a second component that is infertile or otherwise disadvantaged because of some recessive trait (Griffiths and DeLange 1978 Genetics 88:239-254; Perkins 1984 Neurospora Newsl. 31:41-42). am1 ad-3B cyh-1 (FGSC 4564), which is heterokaryon- compatible with both A and a mating types in Oak Ridge background (het-C, -d, -e), has been especially useful. When such a phenotypically wild-type heterokaryon is used as one parent in a cross, the helper nuclei do not participate sexually and all progeny are parented by the disadvantaged component. Heterokaryons with the helper are also useful for stock preservation, assuring survival of genotypes that would otherwise be difficult to maintain. No mutant with an inactive mating type is yet available in the four-spored pseudohomothallic species Neurospora tetrasperma. However, strains containing the dominant gene E (Eight-spore) can serve as helpers. Heterozygous E/E+ crosses are fertile, but homozygous E/E crosses are not, producing barren perithecia with effectively no ascospores (Dodge 1939 J. Hered. 30:467-474). Therefore a marked E strain can be used as a helper in crosses by putting it into a forced heterokaryon with a disadvantaged mutant strain that is E+ and of the same mating type. When such a heterokar- yon is crossed to any E strain of opposite mating type, all progeny will be parented by the E+ component of interest.

An example of the usefulness of E helpers is provided by col(119) (assigned by linkage group VII; Howe and Haysman 1966 Genetics 54:293-302). By itself, the mutant grows slowly as a nonconidiating colony which is difficult to maintain and to cross. In contrast, heterokaryons such as [col(119); pan(124); al(102); E+A + met(123) E A], FGSC No. 7568, and [col(119); pan(124); E+a + lys(112) E a], FGSC No. 7569, are phenotypically wild type and fully fertile when used either as protoperithecial or as fertilizing parent. (The linkage group I marker al(102) is a convenient tag for mating type, with which it does not recombine.) When one of these E + E+ heterokaryons is crossed with an E tester parent of opposite mating type, only the E+ colonial component of the heterokaryon contributes to the progeny. Most ascospores are small and homokaryotic in heterozygous E/ E+ crosses. (A few large heterokaryotic ascospores are produced, but these are readily recognized and avoided when ascospores are isolated manually.) The small single-mating-type ascospores enable haploid genetic analysis to be carried out in conventional fashion, uncomplicated by heterokaryosis in f1 germlings of single-ascospore origin (Calhoun and Howe 1968 Genetics 60:449-459).

Helper heterokaryons have enabled crosses to be made that demonstrate linkage between col(119) and a new morphological mutant lwn (lawn), with about 15% recombination.

4

A Novel Method of Growing Fungi for DNA Extraction

Preparation of fungi for DNA extraction typically involves growing cultures in liquid culture in Erlenmeyer flasks, Roux bottles or even microfuge tubes (Cenis 1992 Nucl. Acids Res. 20:2380). Growing fungal cultures in liquid may require formulating new media or determining aeration requirements, and there are no rapid means of confirming the identification of the resulting mycelium. Fungi are grown routinely on agar media for identification, but agar complicates DNA extraction.

Solutions of 'reverse agar' (BASF pluronic polyol F-127), a block polymer of polyoxypropylene and polyoxyethylene, are solid at normal room temperatures, but liquid at 4C. When the compound is used as a replacement for agar in solid media, its unusual properties allow the separation of mycelium and medium by simply placing a mature culture in a refrigerator. The compound has been employed for isolation of heat sensitive antagonistic microorganisms (Gardner and Jones 1984, J. Gen. Microbiol. 130:731-733; Olson and Lange 1989 Opera Bot. 100:197-199), isolation of enzymes associated with basidiome formation in Coprinus (Choi and Ross 1988 Exp. Mycol. 12:80-83), and for isolating mycelium from Neurospora race tubes (Munkres 1990, Fungal Genet. Newsl. 37:26). In this note, the possibility of extracting DNA suitable for PCR amplification from mycelium grown on reverse agar media is documented.

'Reverse Malt Agar' (RMA) medium was prepared containing 30% BASF pluronic polyol F-127 substituted for agar in the Malt Extract Agar medium of Pitt (1979 The Genus Penicillium, Academic Press). The liquid was poured over the F-127 granules and the resulting suspension left in the refrigerator overnight until dissolved. The solution was then autoclaved, resulting in a thick, sludge like substance, which was again refrigerated, until a homogenous, liquid solution resulted. The liquid was then poured into petri dishes, where it solidified as it warmed to room temperature. The resulting medium is not a true solid, but rather a dense gel.

Cultures of Penicillium spinulosum Thom. (DAOM 216698), Aspergillus japonicus Saito var. japonicus (DAOM 216695), Gliocladium roseum Bainier (Doyle SB-03a, not saved) and Trichoderma harzianum Rifai (DAOM 216501) were grown on RMA for 7 days at 25C in 6 cm petri dishes. Growth rates were slightly slower than on the same medium made with 2% agar, but the resulting colonies produced microscopically typical sporulating structures. For DNA extraction, the petri dishes were placed in the refrigerator for approximately 1 hour until the medium had liquified. Subsequent handling was done on ice. Mycelium was lifted from the medium using an autoclaved pipette tip, placed on the inverted, slanted lid, and allowed to drain for 30-60 seconds. The mycelium was then cut into smaller pieces using a sterile scalpel blade, and transferred into autoclaved 1.5 ml microfuge tubes. The tubes were then spun in a cold microfuge for 5-10 minutes and the excess medium removed. The mycelium was washed twice with 750 uL cold, autoclaved distilled water followed by cold centrifugation, and then used directly for DNA extraction. The DNA miniprep method of Edwards et al. (1991 Nucl. Acids Res. 19:1349), modified by the addition of a cold 70% ethanol wash of the final pellet, was used. The resulting DNA was treated with RNAase A for 1 hour at 37C. PCR amplification of the ITS1-ITS4 region of the ribosomal DNA was performed using the primers and conditions given by White et al. (pp. 282-287 In: PCR Protocols, Innis et al. eds Academic Press). The resulting products were digested for 2 hrs at 37C using HinfI in the buffer supplied with the enzyme.

The DNA yields obtained from mycelium grown on RMA were similar to those obtained from mycelium grown in liquid culture. Washing away excess reverse agar with cold water significantly improved yields, but DNA also was isolated from unwashed mycelium. The resulting DNA performed normally in the ITS PCR amplification and subsequent restriction digests. Use of reverse agar for cultivation of fungi for DNA extraction may be convenient for certain studies. For fungi that do not produce characteristic structures in liquid broth, reverse agar provides a means of ensuring the correct identity of the mycelium before DNA extraction proceeds. Certain population genetics studies, for example, require the manipulation of a large number of cultures that must be cloned (eg. single-spore isolations) before genetic analysis can proceed. The use of reverse agar could eliminate one round of culture transfers, resulting in significant labour savings.

Reverse agar solutions can be stored for some time at 4C and plates poured when required. Experience has shown that poured plates do not keep indefinitely at room temperature. After 4-6 weeks, the medium no longer liquifies, presumably because of higher polymer concentrations resulting from water evaporation. Also, because the polymer itself is slightly inhibitory to fungi, it does not seem to be suitable for weak nutrient media. Our trials with the Fusarium medium SNA (Nirenberg 1981 Can. J. Bot. 59:1599-1609), for example, resulted in sparse growth that could not be harvested following liquification of the medium.

A SODIUM FLUORIDE SENSITIVE MUTANT OF ASPERGILLUS NIDULANS

Fluoride is a widely spread naturally occurring substance in many foods and is used extensively for industrial purposes. The addition of fluoride to drinking water has been assumed to be safe. However, a number of studies have indicated that sodium fluoride is both genotoxic and cytotoxic to mammalian cells (Tsutsui et al. 1984 Mut. Res. 139:193-198). There is conflicting evidence suggesting that NaF is not genotoxic (Kram et al. 1978 Mut. Res. 57:51-55; Martin et al. 1979 Mut. Res. 66:159-167; Li et al. 1987 Mut. Res. 192:191-202) and can suppress the activity of polyfunctional alkylating agents (Obe and Slacik-Erben 1973 Mut. Res. 18:369-371).

Here we report that the alkylating agent-sensitive strains of Aspergillus nidulans L452 (pabaA1;wA3;sagB2), isolated in this laboratory (Swirski et al. 1988 Mut. Res. 193:255-268), also showed an increase in sensitivity to the presence of NaF in the growth medium. In mapping experiments using strain L452 and the derivative strain L455 (yA2;pyroA4;sagB2), it was found that hyphal growth of these strains was completely inhibited by 75 mM NaF. The parent strain L20 (pabaA1;wA3) grew on 75 mM NaF at 50- 60% the rate of the NaF-free controls. L452 and L455 also showed an increased sensitivity to KF. Addition of NaCl or KCl to the growth medium to 100 mM in excess of the normal concentrations had no effect on the growth rates of strains L452, L455, and L20.

The concurrent appearance of an NaF sensitive phenotype in the alkylation repair defective strains of A. nidulans suggested that the effect might be due to the presence of the sagB2 mutations. The sensitivity to NaF and KF of L452 and L455 is not a pleiotropic effect of the sagB2 mutation; this was demonstrated by the separation of the two phenotypes in crosses (R.F.=27-47%). Haploidization studies showed that the mutation which confers sensitivity to NaF is located on chromosome VIII: sagB2 is also located on chromosome VIII. The sensitivity to NaF and KF of these strains is functionally recessive to the wild-type.

Other notable NaF sensitive mutants of A. nidulans are the palB strains (e.g. G832 - galH7;facB101 riboB2 palB7 chaA1); the sensitivity in this case is due to a defective alkaline monophosphoesterase activity (Dorn 1965 Genet. Res. Camb. 6:13- 26). Our strains showed wild-type acid and alkaline monophosphoesterase activities as judged by histological staining methods. Unlike the palB strains, the internal pH of L20, L452, and L455 was found to be wild type using the method of Caddick et al. 1984 Mol. Gen. Genet. 203:346-353. The wild-type NaF response of an L455::G832 diploid strain provided further evidence that our mutation was not a variant of the palB locus mutation. Furthermore, both L452 and L455 showed a wild-type response to X-irradiation in an oxygen-free environment. This suggests that the mutation conferring NaF sensitivity on these strains is unlikely to cause a defect in the radiation response/repair systems.

In conclusion, we have revealed a functionally recessive mutation on chromosome VIII of the A. nidulans strain L452 which results in an increased sensitivity to NaF. The data show that this mutation probably arose by an independent event during the mutagenesis procedure used to induce the sagB2 mutation. This new locus has been designated nfsA; its function is as yet unknown.

DETECTION OF SER/THR PROTEIN PHOSPHATASES IN NEUROSPORA CRASSA

Protein phosphorylation is a frequent posttranslational modification regulating cellular processes in eukaryotes. The phosphate content of a protein is determined by the conflicting activities of protein kinases and phosphatases. Protein phosphatases were divided into Ser/Thr and Tyr specific groups, depending on the phosphorylated residue in the substrate molecules. The former group was further classified based on enzymatic criteria (reviewed in Cohen 1989 Ann. Rev. Biochem. 58:453-508). Protein phosphatase 1 (PP1) is inhibited by two heat stable proteins termed inhibitor-1 and -2. Protein phosphatase 2A is inhibited by nanomolar concentration of the tumor promoter okadaic acid. Protein phosphatase 2B (PP2B) - also called calcineurin - is stimulated by Ca-calmodulin, and protein phosphatase 2C (PP2C) is a Mg2+ dependent enzyme. Molecular cloning of the catalytic subunits revealed that PP1-PP2A-PP2B consist of a highly conserved superfamily of proteins.

The catalytic subunit of PP2B from N. crassa was cloned and expressed (Higuchi et al. 1991 J. Biol. Chem. 266:18104-18112). We assumed that PP1 and PP2A were also present in this organisms. In order to test this possibility we assayed phosphatase activities in N. crassa. For the detection of PP1 and PP2A we adapted a method originally described for S. cerevisiae (Cohen et al. 1989 FEBS Lett. 250:601-606). PP2B activity was measured according to Higuchi et al. (1991 J. Biol. Chem. 266:18104-18112).

Neurospora crassa strain RL3-8 (FGSC stock number 2218) was cultured in Vogel's minimal medium for 48 h at 27oC. Two agar slants (one week old) were used to inoculate 400 ml of medium. Mycelia were harvested by filtering through cheesecloth. The mycelial mat was weighed, frozen in liquid nitrogen and stored at -70oC. Homogenization was carried out by disintegrating the mat at the temperature of the liquid nitrogen in mortar with pestle, then by mixing the powder with 4 volumes/weight extraction buffer (50 mM Tris-HCl pH 7.4, 1 mM EDTA, 0.1% vol/vol. beta-mercaptoethanol) containing a mixture of freshly diluted protease inhibitors (5 mM benzamidine, 0.2 mM phenylmethyl sulfonyl fluoride, 1 mM o-phenanthroline). Extraction was completed by mixing the ice cold homogenate at full speed for 1 min in an MSE blender. The extract was filtered through glass wool and was centrifuged at 4oC and 15,000 g for 15 min. The supernatant was used as crude extract in the phosphatase assays. We found that small aliquots (0.1-0.5 ml) of extract can be stored, after being frozen in liquid nitrogen, at -70oC for 7 days without significant loss of phosphatase activity.

PP1 and PP2A were assayed with rabbit muscle phosphorylase a substrate. Phosphorylase a was prepared from phosphorylase b with [gamma-32P]ATP and phosphorylase kinase. The assay mixture contained 5 uM 32P-phosphorylase a dimer and 5 mM caffeine in the extraction buffer in a total volume of 30 ul. The reaction was initiated by the addition of the substrate and was terminated after 10 min incubation at 30oC by the addition of 100 ul 10% TCA. TCA soluble radioactivity was counted by Cherenkov radiation. Maximal specific activity was obtained if the extract was diluted with extraction buffer, so that less than 20% of the total 32P was released from the substrate. To test if the liberated radioactivity was in Pi or small phosphopeptide(s) we converted Pi into phosphomolybdic acid and extracted it with organic solvents (Antoniw and Cohen 1976 Eur. J. Biochem. 68:45-54). More than 95% of the radioactivity was recovered in the organic phase. In addition, 20 mM NaF (a well known phosphatase inhibitor) reduced the activity to 5-8%. These facts demonstrate that the assay measures phosphatase activity and the effect of proteases can be neglected.

After establishing the assay system we investigated the effect of specific phosphatase inhibitors. One micromolar okadaic acid caused a nearly complete inhibition with the half maximal effect at 10 nM (Fig.1.). Since the IC50 of okadaic acid is 0.1 nM for PP2A and 15- 20 nM for PP1 in mammalian tissue extracts (Cohen et al. 1989 FEBS Lett. 250:596-600) we assumed that most of the phosphorylase phosphatase activity in N. crassa extract was due to PP1. To test this hypothesis we measured the effect of okadaic acid in the presence of an excess of inhibitor-2. As shown in Fig. 1, under these conditions the activity was half maximally inhibited by 0.3 nM okadaic acid and 90% inhibition was observed at 1 nM concentration. The sensitivity of N. crassa PP1 to the rabbit muscle inhibitor-2 was demonstrated in Fig.2. When PP2A was inactivated with 1 nM okadaic acid, 10,000 U/ml inhibitor-2 caused complete inhibition. 50% inhibition was observed at 30 U/ml concentration i.e. around 1 U inhibitor-2 in the 30 æl assay mixture elicited half maximal inhibition.

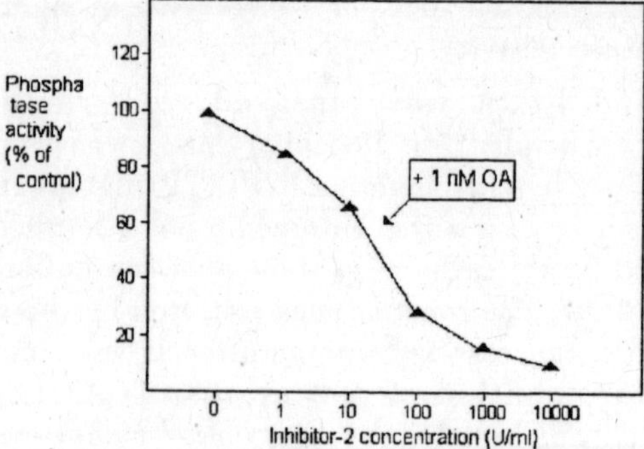

Figure. Inhibition of phosphorylase phosphatase activity by okadaic acid in the presence and absence of inhibitor-2 (I-2)

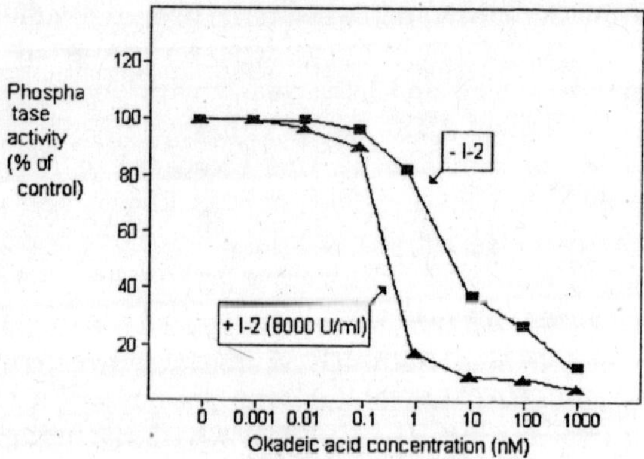

Figure. Inhibition of phosphorylase phosphatase activity by inhibitor-2 in the presence of okadaic acid (OA)

Our results indicate that the phosphorylase phosphatase activity inhibited by 1 nM okadaic acid can be attributed to PP2A and the one inhibited by 10,000 U/ml inhibitor-2 belongs to PP1. Protein concentration was measured according to Read and Northcote (1981 Anal. Biochem. 116:53-64). Using these methods we found 4.95 +/- 0.9 mU/mg protein (n=10) of total phosphorylase phosphatase activity in N. crassa mycelial extract, 20% of which was PP2A and 65% PP1. By definition one unit of the enzyme liberated 1 umol Pi in 1 min under the above conditions.

PP2B was assayed with rabbit muscle inhibitor-1. The substrate was phosphorylated with [gamma-32P]ATP and bovine cAMP-dependent protein kinase. The reaction mixture contained 3 uM 32P-inhibitor-1, 40 mM Tris/HCl pH 7.0, 0.4 mM DTT, 0.2 mM CaCl2, 1 mM MnCl2 and 12 uM calmodulin in 30 ul total volume. The reaction was started by the addition of substrate and after 10 min incubation at 30oC it was terminated by the addition of 100 ul 20% TCA and 100 ul 6% BSA. The released radioactivity was counted as above. Using the extraction method, we proved that more than 95% of the TCA soluble radioactivity was due to the presence of 32Pi. About 40% of the total activity was inhibited by 0.5 mM EDTA. The activity stimulated by Ca2+-calmodulin in the presence of Mn2+ was attributed to PP2B, thus the specific activity of PP2B was estimated to be 0.044 +/- 0.01 U/mg protein (n=6) in the extracts. One unit of PP2B liberated 1 nmol Pi in 1 min in the above assay. The assays described in the present communication are suitable to test protein phosphatase mutants by measuring crude N. crassa extracts, and provide a handle for the purification and characterization of the respective enzymes. The results show considerable similarity between the N. crassa and animal protein phosphatases suggesting that cloning of PP1 and PP2A catalytic subunits could be possible using heterologous cDNA probes.

MAPPING OF MUTANTS RESISTANT TO P-FLUOROPHENYLALANINE IN DIPLOID *ASPERGILLUS NIDULANS*, LETHAL IN HAPLOIDS

In a previous paper (Babudri and Morpurgo 1990 Curr. Genet. 17:519-522) we described a new class of para-fluorophenylalanine (FPA) resistant mutants in *Aspergillus nidulans*. These mutants were obtained by plating UV irradiated diploid conidia on minimal medium (MM) supplemented with FPA (0.188 mg/ml). Some of them have characteristics different from those previously described, showing resistance to FPA in the diploid strains and being lethal in hemizygous or homozygous condition, either in presence or absence of FPA. This means that the genes involved affect fundamental processes of the fungus and at the same time, can mutate to give FPA resistance.

The diploid strain we used was not marked on all eight chromosomes of *A. nidulans* and it was impossible to map all the lethal mutants; we have therefore synthesized a new diploid using our strain 35 (*anA1 pabaA1 yA2;sC12;methG1;nicA2*) and the tester strain FGSC 513 (*adE20;AcrA1;ActA1;pyroA3;facA303;lacA1 sB3;choA1;chaA1*) which allows precise chromosomal mapping. We irradiated the conidia with weak doses of UV rays and selected diploid resistant mutants on MM plus FPA. These were haploidized by seeding conidia on MM supplemented with sublethal doses of Benomyl. The microcolonies growing in the Benomyl supplemented medium were transferred on complete medium (CM) and all the segregant sectors were analyzed, testing the FPA resistance. Ten out of thirteen selected produced rare FPA resistant mutants, presumably haploid, all requiring pyridoxine and methionine (*methG1* and *pyroA3* map on chromosome IV).

These mutants were not further analyzed because it is possible that they have an impaired capacity for uptake of amino acids as do some FPA resistant mutants isolated by Sinha (1967 Genet. Res. 10:261-272) and Tiwary et al. (1987 Curr. Microbiol. 15:305-311). The mutants 302-11, 302-12 and 302-15 failed to produce haploid FPA resistant segregants even if the colonies grown on the Benomyl supplemented medium were transferred to CM supplemented with FPA on which the resistant segregants could be selected. The analysis of the FPA sensitive sectors gave the following results: 302-11 and 302-15 map to chromosome VIII.

The situation of 302-12 is more complicated. The diploid FPA resistant strain does not utilize lactose and requires thiosulphate and is therefore a non-disjunctional diploid homozygous for chromosome VI. The analysis of the sectors mapped the gene on chromosome IV; all the sectors required methionine and this excludes the possibility that the mutant has impaired uptake of amino acids. It is not known if the requirement for thiosulphate is involved in FPA resistance but it is noteworthy that a mutant diploid described in the previous paper is thiosulphate requiring. Considering these results and those obtained in the previous work, three loci mapping to chromosomes I, IV and VIII which determine high level FPA resistance and are lethal in haploids have been identified. Research is in progress to determine 1) how many different loci may exhibit these characteristics and 2) the physiological basis of the phenomenon. This second point may be achieved by looking for temperature sensitive conditional lethals.

In haploids the segregation of the chromosomes should be random. The mapping of FPAR lethals is determined by the absence of a chromosome in a translocation free diploid strain. In the strain 302-11, chromosome VIIIb was missing in the 32 haploid segregants tested. The same chromosome was missing from the 37 haploid sectors tested in the strain 302-15. In the strain 302-12, all the haploid sectors lacked chromosome IVb. The FPAR mutant nøl (see Babudri and Morpurgo, 1990) was obtained in a different diploid but the method used for mapping was the same; this mutant was mapped on chromosome I.

LONG TERM STORAGE OF *PODOSPORA ANSERINA*

Until now, maintenance of *Podospora anserina* strains was tedious, requiring yearly sexual crosses, spore isolation and germination, verification of the genotype and transfer of mycelium to stock tubes maintained at 4 C. We have circumvented these difficulties by storage of ascospores at -80 C. Sexual crosses were made on synthetic medium, either by confrontation or spermatization, to reach high fertility. When perithecia began to shoot ascospores, plates containing agar/salt medium were put upside down over plates containing perithecia. Ascospores were recovered after 24 h by scraping the surface of the agar with a sterile spatula. More than 10(4) asci were recovered from each plate. They were transferred from the spatula to a 1.2 ml Nalgene cryovial containing 0.3 ml sterile storage medium (3 parts 50% glycerol: 7 parts liquid corn meal medium). The vials were vortexed 30 sec and placed at -80 C. After two years storage, the germination rate of ascospores was only slightly lower than that of freshly isolated ones and several cycles of freeze-thawing did not decrease it significantly. This method has been successfully applied to wild-type strains of *Podospora anserina* and to mutant strains displaying ascospores of wild-type phenotype. We are now testing with strains which have more fragile ascospores. For methods and media used in growing and crossing *Podospora anserina*, see Esser, K. 1969. Neurospora Newsletter 15:27-31.

Table. Survival of wild-type ascospores after -80 C storage

	Time of storage			
strain*	**2 months**	**14 months**	**24 months**	**24 months+**
s	20/20	60/60	52/87	71/92
S	8/8	nd	77/83	nd
A	17/17	79/80	nd	72/82

* : s, S and A are three wild-type strains of Podospora

+ : ascospores after 3 cycles of freeze-thawing

CBM1, A *NEUROSPORA CRASSA* GENOMIC COSMID LIBRARY IN PAC3 AND ITS USE FOR WALKING ON THE RIGHT ARM OF LINKAGE GROUP VII

Gene cloning in *Neurospora crassa* is often achieved by mutant complementation. However, the cloning strategy sometimes requires the isolation of a specific genomic region (by chromosome walking) before transformation of *N. crassa* . This is the case, for example, if the gene to be isolated has a non-selectable phenotype. Here we specifically describe the construction of the cosmid vector, pAC3, which is designed for direct transformation of *N. crassa*, its utilization for the construction of a genomic library, and chromosome walking in the region of *un-10* on linkage group VII.

EXPERIMENTAL PROCEDURES

The pAC3 cosmid vector was constructed as follows:

(a) The prokaryotic origin of replication, the antibiotic marker and cos sequences were derived from pJB8 (Ish-Horowicz et al. 1981. Nucl. Acids Res. 9:2989).

(b) The pJB8 EcoRI fragment containing the *Bam*HI cloning site was replaced by the *Eco*RI-*Eco*RI segment from pWE15 (Catalogue no. 851201 from Stratagene). This 500 bp segment contains two *Not*I sites and the T3 and T7 phage promoters flanking the *Bam*HI cloning site.

(c) A *Bam*HI-*Bgl*II fragment from the plasmid MSK338, containing the Neurospora *qa-2* gene (coding for catabolic dehydroquinase) was treated with Klenow DNA polymerase 1 to fill in the protruding ends. Such a fragment was blunt ligated in the filled-in *Sal*I site of the construct obtained in b). The final product is the cosmid vector pAC3.

We used the pAC3 vector to construct the CBM1 *N. crassa* genomic library. We isolated partially digested chromosomal DNA directly from 74A *N. crassa* agarose plugs prepared as for pulse field electrophoresis (Ballario et al. 1989. Fungal Genetics Newsletter 36:38). The chromosomal DNA was dephosphorylated and ligated at a pAC3 arms-insert ratio of 3:1 (see below for the definition of arms). To avoid the loss of methylated sequences (Grant et al. 1990. Proc. Natl. Acad. Sci. 87:4645) and the recombination events between repeated sequences, a Gigapack Gold packaging extract (Catalogue no. 200214 from Stratagene) was used for packaging and *E. coli* PLK-F' cells (*recA lac mcrA mcrB hsdR gal supE [F' proAB lacqZ M15 Tn10 (tetr)*]) were infected. An efficiency of about 1.2 x 10(5) clones per ug of *N. crassa* chromosomal DNA was obtained.

RESULTS AND DISCUSSION

The characteristics of pAC3 are summarized as follows:

1) The cosmid pAC3 allows the use of the strategy suggested by Ish-Horowicz and Burke for cloning in pJB8. In fact, the presence in pAC3 of asymmetric sites (like *Hin*dIII and *Pst*I) enables the preparation of dephosphorylated left and right cos fragments (defined arms). The chromosomal DNA, partially digested by *Sau*3A or *Mbo*I, is dephosphorylated and used without a previous step of size enrichment. Before ligation, the cos fragments are digested with *Bam*HI to create the phosphorylated

cloning site. This cloning procedure allows one to avoid two potential problems: a) the formation of cosmids containing inserts with non-contiguous genomic fragments, b) the formation of cosmids containing multiple copies of the vector.

2) The origin of replication of pJB8, present in pAC3, maintains the cosmid at a low copy number and may permit survival of clones containing sequences toxic for *E. coli.*
3) The presence of T3 and T7 phage promoters flanking the cloning site allows both the preparation of RNA probes representing the ends of the cosmids and restriction mapping using T3 and T7 DNA sequencing primers (Evans et al. 1989. Gene 79:9).
4) The sites for a rare cutter restriction enzyme (*Not*I) flanking the cloning site make it possible to remove the inserts and facilitate physical restriction mapping.
5) A selectable marker for direct selection of *N. crassa* transformants has been included.

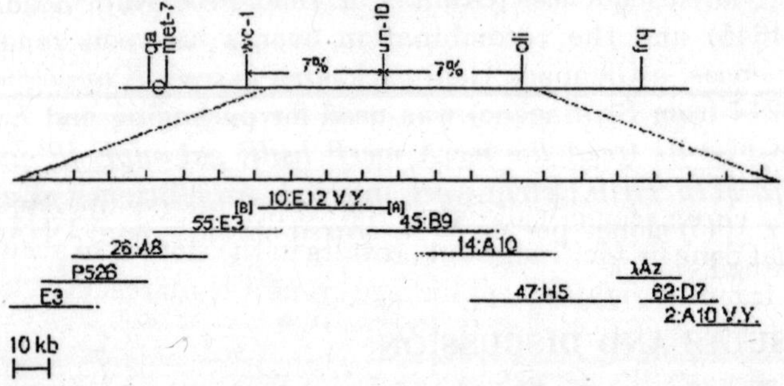

Fig. Chromosome walking from *un-10* toward *wc-1* and *oli*. Part of the linkage groups VII map is shown on the top, the genetic distances in the region between *wc-1* and *oli* are indicated. The inserts of the walking cosmids and lambda clone are shown as horizontal lines below the chromosome. The cosmids indicated with the symbol V.Y. are from the Vollmer-Yanofsky library, the others are from CBM1. Probes A and B, used to start the walks, are indicated by boxes. Cosmid 10:E12 was kindly provided by T. Schmidhauser. The clone Az was isolated from the lambda J1 library (M. Orbach) available from FGSC.

We have used the CBM1 library for a chromosome walk from *un-10* toward white collar 1 (*wc-1*) on the right arm of linkage group VII. As a starting point for the walk, the cosmid clone 10:E12 from the Vollmer-Yanofsky (V.Y.) library complementing the *un-10* mutation was used. In the absence of indications about the genomic orientation of cosmid 10:E12, we were forced to walk in both directions. Fragments A and B (in Fig. 2) of cosmid 10:E12 were identified by conventional restriction mapping and Southern hybridization and used as starting points in the walk.

Starting from probe A we made four steps of walking, covering about 120 kb, and finally found the cosmid 62:D7 from CBM1 overlapping the 2:A10 cosmid from the V.Y. library. This clone was previously mapped at the extreme left of *oli* by McClung et al. (1989. Nature 339:558) and revealed the walking orientation toward *frq*. The isolation of the last cosmid (62:D7) was preceded by a walking step in a Neurospora lambda library to isolate the clone Az. This step was necessary to cover about 4 kb apparently absent in both the V.Y. library and ours.

Starting from the probe B we have made four steps so far, representing about 80 kb of genomic sequences in the direction of wc-1. All the walking steps have been checked with the RFLP mapping technique (done by A. Folgori) always confirming the assignment of the selected cosmids to the correct linkage group VII region.

This chromosome walking in the CBM1 library, together with that done by McClung et al., results in the cloning of about 390 kb from the right arm of linkage group VII represented by a series of contiguous cosmids. Having this large region already cloned could allow the molecular investigation of the several known genetic markers mapped in the region.

THE NOMENCLATURE OF *MTS* AND *CPC* MUTANTS OF NEUROSPORA

Mutants defective in general amino acid control in Neurospora have been designated *cpc* for cross pathway control (Barthelmess 1982, Genet. Res. 39:169-185, Davis 1979, Genetics 93:557-575). More recently, Koch and Barthelmess (1987, FGN 33:30-32, 1988, FGN 35:22-23) have shown that mutants sensitive to 5-methyl tryptophan and designated *mts* are *cpc-1* alleles. Despite the priority of *mts*, on the grounds of apt description, I propose that the designation *mts* lapse and *cpc* be used for this set of alleles.

Mutants of Neurospora supersensitive to 5-methyl tryptophan (5MT) were isolated as a tool to assist selection of mutants defective in the regulation of tryptophan biosynthesis (Catcheside 1966, PhD Thesis, University of Birmingham). It had been found that Neurospora was resistant to high concentrations of tryptophan analogues, including saturated solutions of 5MT, and it was reasoned that if sensitive mutants could be obtained they should permit selection of mutants defective in tryptophan pathway control as second site mutations restoring resistance.

Twelve mutants sensitive to 5MT were obtained from a single experiment. Eleven of these, MN1 to MN8, MN10, MN13 and MN16, were tentatively assigned to a single locus, *mts*, 0.5 units distal to *ylo-1* on linkage group VI, on the basis that each recombined with a frequency of 0.03% or less with MN1. Crosses between MN9 (three strains) and MN1 (two strains) yielded 2.1%, 3.7% and 5.1% MT resistant progeny. Koch and Barthelmess (1988) reported that MN9 fails to complement either MN1 or *cpc-1* alleles, indicating each of these three mutations to be allelic despite the relatively high recombination frequency between MN1 and MN9. Koch and Barthelmess also reported that MN9 contains a duplication generating chromosomal rearrangement involving linkage groups IV and VI.

Barthelmess and Krüger (unpublished) investigated the expression of *cpc-1* in MN9. When total RNA of MN9, wild-type and *cpc-1* mutant strains was investigated in Northern blots, no hybridization with a radiolabelled probe of the *cpc-1* gene was found for MN9. This would indicate that the *cpc-1* gene is involved in the rearrangement of MN9. In view of this, all of the original 5MT resistant mutants were rechecked. Crosses to Emerson wild types show the aberration associated with MN9 to be present in the original isolate and all tested progeny of backcrosses to wild type. MN3, MN4 and MN5 also yield unpigmented spores in crosses to wild type, indicating they too contain chromosomal rearrangements. However, there is no evidence of a chromosomal rearrangement in the remaining nine MN alleles. Information on the plieotropic phenotype of MN1 has been published (Catcheside 1971, Proc. Austral. Biochem. Soc. 4:17, 1978, Neurospora Newsl. 25:17-18). The enhanced sensitivity of MN1 to 8-aza-adenine suggests that cross pathway control may extend beyond thc regulation of amino acid pathways in Neurospora.

A SIMPLE, RAPID PROCEDURE FOR THE ISOLATION OF DNA FOR PCR FROM *GIBBERELLA FUJIKUROI* (*FUSARIUM* SECTION *LISEOLA*)

The polymerase chain reaction (PCR) is a method for amplifying specific segments of DNA defined by the small primers used to start the reaction. Using arbitrarily chosen 10-base primers, one can generate "random amplified polymorphic DNA" (RAPD) markers (Williams et al. 1991 Nucl. Acids Res. 18:6531-6535). These DNA fragments, separated by electrophoresis in an agarose gel, can be used as markers for studying genetic variation within and among fungal populations. A rapid and simple procedure for isolating fungal DNA from multiple isolates is needed if this technique is to be useful for diagnosis and screening of natural populations. We have developed a method for isolating DNA without grinding from Fusarium cultures grown on agar slants using a modification of the SDS miniprep method of Lee et al. (1988 Fungal Genetics Newsletter 35:23-24).

1. Place a cube (0.5 cc) of mycelia-covered agar from a 5 day-old slant in a 1.5 ml microcentrifuge tube.
2. Fill tube with liquid nitrogen and let it evaporate. Repeat. No grinding is necessary.
3. Immediately add 500 ul of 65 C lysis buffer (50 mM Tris pH 8, 50 mM EDTA, 3% SDS, 1% BME, and 0.1 mg/ml Proteinase K). Vortex tube and place in 65 C water bath for 1 hr. Vortex after 30 min and 60 min.
4. Extract with 500 ul phenol. Spin tube 5 min at 8000 rpm in microcentrifuge to separate phases. Remove 450 ul of the aqueous phase.
5. Extract with 450 ul buffered phenol. Spin. Remove 400 ul of the aqueous phase. 6. Extract with 400 ul

chloroform:isoamyl alcohol::24:1. Spin. Remove 350 ul of the aqueous phase.

7. Add 50 ul 7.5 M ammonium acetate. Gently mix. Add 880 ul 95% ethanol. Invert to mix. Place in -20 C freezer 30 min to overnight.
8. Spin down DNA pellet for 20 min at 13,000 rpm. Rinse pellet in 70% ethanol. Dry pellet and resuspend in 20 ul TE (10 mM Tris, 1 mM EDTA, pH 8).

DNA was prepared from isolates grown on complete and minimal slants (Correll et al. 1987 Phytopathology 77:1640-1646). We obtained enough DNA from each isolate for 20 reactions. We used 1 ul of DNA for PCR with our primer, designated ECORI (5'- ATGAATTCGC-3'). To each reaction tube on ice was added 42 ul sterile, glass-distilled and deionized water; 5 ul 10X buffer from Promega (500 mM KCl, 100 mM Tris-HCl pH 9, 15 mM MgCl, 0.1% gelatin w/v, 1% Triton X-100), 1 ul of 50 uM primer, 1 ul of 1 mM dNTP's and 1 ul DNA solution. Tubes were boiled for 2.5 min. Tubes were returned to ice, then 1 unit of Taq polymerase (Promega) was added. Tubes were spun for 5 sec in a 4 C microfuge and then 100 ul mineral oil was added.

Tubes were placed in a PTC-100 programmable thermal controller from M.J. Research with the following program: Step 1. 92 C 30 sec. Step 2. 35 C 1 min. Step 3. Slope 35 to 72 at 1 degree C every 8 sec. Step 4. 72 C 2 min. Step 5. cycle to Step 1. 45 times. Step 6. 72 C 7 min. and Step 7. end. 20 ul of each reaction mix was loaded into a 1.3% agarose gel with TBE buffer. After 2.0 hours at 56 V, the gel was stained with ethidium bromide and photographed. The resulting patterns. of amplified DNA fragments are listed in Table below.

Table. RAPD marker patterns for Gibberella fujikuroi isolates using the ECORI primer

Strain	DNA fragments, kb +/- SE
A2910	1.661 +/- .044, 0.597 +/- .015
A2911	1.698 +/- .098, 0.561 +/- .031
A3957	1.678 +/- .032, 0.639 +/- .015
X3974	1.661 +/- .044

CONSTRUCTION OF TWO *NEUROSPORA CRASSA* LAMBDA ZAPII-CDNA LIBRARIES

In the presence of the anti-microtubule agent benomyl, *N. crassa* synthesizes and secretes a major 60 kDa protein in the culture medium. This protein is absent in untreated cells (Rossier et al. 1989. Eur. J. Cell Biol. 50:333-339). To clone the gene coding for the 60 kDa protein and to unravel its function, we have constructed a cDNA library from benomyl-treated cells ("benomyl" cDNA library). A second cDNA library was constructed from untreated cells in order to clone genes coding for cytoskeletal proteins ("standard" cDNA library).

The two cDNA libraries have been constructed using the Lambda Uni-Zap II vector in the ZapTM cDNA synthesis kit (Stratagene, USA; Short et al. 1988. Nucleic Acids Res. 16:7583-7600). The advantage of this cloning vector is that each cDNA is in the proper orientation to facilitate expression of fusion proteins by the *lacZ* promoter in the Lambda Zap II vector for antibody screening. cDNA clones can also be identified by hybridization with a DNA probe. The pBluescript SK(-) plasmids containing the cDNA inserts can be excised from the purified Lambda Zap II clones with filamentous helper phage R408 for immediate restriction enzyme mapping, sequence analysis and in vitro mutagenesis. The pBluescript SK(-) plasmid is a 2958 base pair phagemid derived from pUC19 and carries an fl origin of replication allowing single strand cDNA rescue, via filamentous helper phage infection, for site-specific mutagenesis or single strand sequencing. It possesses a *colE1* origin for replication in *E. coli* and the *AmpR* gene for selection on ampicillin medium.

The "standard" cDNA library has been made with polyA+ mRNAs from *N. crassa* (wild type, St. Lawrence 74A) cultivated in liquid Vogel medium for 19 hours and the "benomyl" cDNA library from *N. crassa* treated for 8 hours with 1.7 uM benomyl after 17 hours of growth in liquid Vogel medium. Total RNA from *N. crassa* was isolated by sedimentation through a CsCl gradient, solubilization in guanidine-HCl at 65 C and ethanol precipitation (Hoang-Van et al. 1989. Eur. J. Cell Biol. 49:42-47). 1.25 ug of poly A+ mRNAs were purified on oligo(dT)-cellulose columns and used as templates to construct the two cDNA libraries. First strand synthesis was primed with a poly(dT)-primer/linker containing a *Xho*I restriction site and performed with Moloney reverse transcriptase in the presence of 5-methyl-dCTP.

The cDNAs were blunt ended using T4 DNA polymerase and ligated to *Eco*RI adaptors. These *Xho*I-*Eco*RI cDNAs were then unidirectionally inserted into the Lambda Zap II vector (1 æg), which has been predigested with *Xho*I and *Eco*RI and dephosphorylated with alkaline phosphatase. The 5' end of the mRNAs is then nearest to the *lacZ* promoter. The two *N. crassa* cDNA libraries were packaged with Gigapack Gold packaging extracts (Stratagene, USA). A total of 1.1 million recombinant phages for each library were obtained after the initial infection of PLK-F', a *mcrA*-, *mcrB*- *E. coli* strain, preventing digestion of hemi-methylated cDNA. Recombinant phages may be recognized by blue/white color identification when plated on lac- hosts in the presence of isopropyl- -D-thio-galacto-pyranoside (IPTG) and 5-bromo-4-chloro-3- indolyl- -D-galacto-pyranoside (X-Gal). A blue to white of <0.5% for the "standard" cDNA library and <1% for the "benomyl" cDNA library indicated that most of the resulting phages were recombinant. The titers before amplification of the "standard" and "benomyl" cDNA libraries were 1.5 x 109 and 109 plaques/ml, respectively.

We have screened the fusion proteins from the "benomyl" cDNA library with rabbit polyclonal antibodies raised against the 60 kDa protein secreted by N. crassa in the presence of 1.7 uM benomyl. Fusion proteins expressed in the presence of 10 mM IPTG and then adsorbed onto nitro-cellulose filters were incubated with the anti-60 kDa antibodies (dilution 1:2000 in TBS containing 0.5% Tween-20).

The peroxidase activity was revealed with 0.07 mg/ml diaminobenzidine-tetrahydro- chloride (DAB) in the presence of 0.03% H2O2 after incubation of the nitrocellulose filter with secondary antibodies (horseradish peroxidase-labeled donkey anti-rabbit IgG, Amersham, UK) (dilution 1:2000 in TBS containing 0.5% Tween-20). 175 positive clones were obtained after screening 280,000 plaques. 9 positive clones with insert of various sizes were analyzed by restriction enzyme mapping. Their sizes ranged from about 0.5 to 1.6 kb. Their fusion proteins overexpressed in the presence of 10 mM IPTG are shown in Fig. below. The molecular weights of the fusion proteins were in agreement with the sizes of the inserts. Clones N65 and N121 (possibly full length cDNA clones) were inhibitory for *E. coli* growth (results not shown) and did not overexpress the fusion proteins in the presence of IPTG.

The *N. crassa* beta-tubulin cDNA has also been cloned by screening fusion proteins of the "standard" cDNA library with rabbit polyclonal antibodies raised against *N. crassa* beta-tubulin and monoclonal antibodies raised against chick brain beta-tubulin (Amersham, UK) (Hoang-Van, preparation).

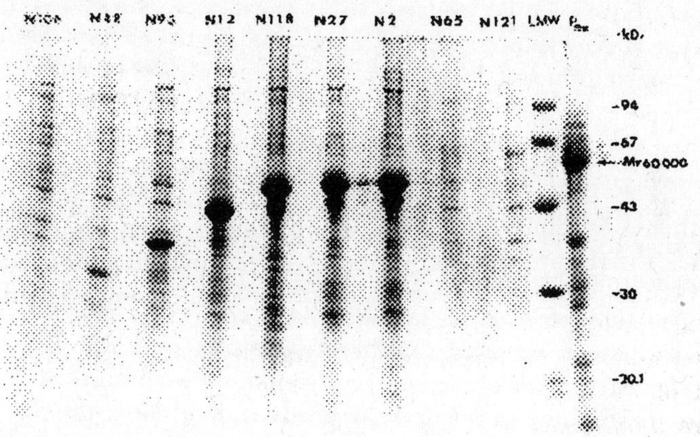

Figure. Fusion proteins overexpressed in E. coli by cDNA inserts of various sizes of the gene coding for the 60 kDa protein secreted by benomyl-treated cells of N. crassa. Overexpression was induced in the presence of 10 mM IPTG. These fusion proteins in the form of inclusion bodies were purified according to Marston (1986. Biochem. J. 240:1-12), dissolved in 8 M urea and separated by SDS-PAGE (10% acrylamide). LMW: Low molecular weight standards. Pex: extracellular 60 kDa protein (arrow) from culture medium

FILTRATION-ENRICHMENT METHODS FOR SELECTING AUXOTROPHS AND OTHER MUTANTS IN *ASCOBOLUS IMMERSUS* AND SIMILAR FILAMENTOUS FUNGI.

B.C. Lamb and S. Helmi(2)- Department of Biology, Imperial College of Science, Technology and Medicine, London SW7 2BB, United Kingdom. (2)Current address: Dept. of Environmental Studies, Institute of Graduate Studies and Research, University of Alexandria, Egypt. Filtration-enrichment is a very successful method for obtaining auxotrophs, but traditional methods cannot be applied directly to many fungi lacking suitable asexual spores, mainly because of a heavy carry-over of parental nutrients into sexual spores. A procedure has been devised to apply filtration-enrichment to such fungi by using hyphal mutagenesis, crossing, germination of sexual spores, fragmentation of germination hyphae into small propagules, and then two cycles of filtration-enrichment in liquid minimal medium (LMM). It has been optimized in *Ascobolus immersus*, Pasadena strains, for mutagenesis, hyphal fragmentation and recovery, filter pore size, duration and number of growth periods, and final isolation of colonies. Many variations on these factors were tried. The optimum conditions described here gave yields of about 30% of auxotrophs. The method is easily applicable to other fungi, such as Sordaria, whether homothallic of heterothallic. It could also be used in fungi where the asexual spores are large, with a heavy carry-over of parental nutrients, as in Cochliobolus, and in those Fusarium species which have sexual reproduction.

The ascospores of *A. immersus* are about 55 um x 33 um and so cannot be used for direct selection in LMM as the massive carry-over of parental nutrients allows a mutant (e.g. lysine-requiring) ascospore to produce hyphae growing in unsupplemented LMM almost as fast as wild-type ones, giving dense colonies up to 20 mm diameter. Lamb (1982 Micr. Gen. Bull. 52:2-4) suggested reducing this carry-over by germinating the ascospores, then fragmenting the hyphae in a blender into lengths of 1-3 coenocytic cells, when the terminal septa will usually seal in the contents, and the fragments can regenerate hyphae and colonies. The short healed fragments can then be used as if they were germinated conidia in selective methods, with a nutrient carry-over about two orders of magnitude less than in ascospores. Mutagenizing the hyphae used in a cross allows segregation-delay to occur before meiosis, so that each haploid sexual spore will either be wholly wild-type, or wholly mutant, even if not uninucleate. The present methods are based on those of Lamb (1982, loc. cit.), but with various practical improvements which more than tripled the yield of auxotrophs.

Strains, media and general methods were as described by Yu-Sun (1966 Genetica 37:569-580) and Helmi and Lamb (1983

Genetics 104:23-40). Crosses were made in the light at 17.5 C and growth was at about 24-26 C. Nylon filters of defined pore size were monofilament bolting cloth (Henry Simon Ltd., Stockport, UK). For mutagenesis and crossing, three squares of cellophane, 2.5 cm square, were spread on crossing medium in 9 cm diameter petri dishes. A small inoculum of (-) mating type was placed centrally on each square and allowed to grow for two days, then the original agar inoculum block was removed, leaving a thin layer of fully exposed hyphae. For chemical mutagenesis, the cellophane (plus hyphae) was removed and immersed in mutagen solution, then washed several times in LMM. For UV mutagenesis, the hyphae were kept in the dark for four hours after irradiation. At the same time as (-) inoculation, a (+) mating type strain was inoculated at one edge of petri dishes of crossing medium. After mutagenesis, the cellophane squares with treated (-) hyphae were placed, hyphae downwards, on the opposite side of the dishes to the non-mutagenized (+) inoculum. The cellophane was removed about 4 hours later, leaving the (-) hyphae on the agar. These crosses were left at 17.5 C until ready. The efficiency of mutation was scored from the frequency of 4w+(red ascospores):4w (white ascospores) segregations in dehisced octads of ascospores in w+ x w+ crosses. The optimum mutagenic treatments to use on one strain in a cross were: UV, 240 secs at 800 ergs/sec/c2; ICR-170, 10 ug/ml for 2700 secs, NMG, 50 ug/ml for 600 secs. These gave, respectively, 4.9, 4.9 and 2.2% of 4w+:4w octads from samples of about 12,000 asci, compared with spontaneous frequencies of about 0.1%.

For filtration-enrichment, dehisced ascospores were separated from each other with 5 ml of 0.1% w/v pronase solution per collecting lid for 30 min. Red ascospores from w+(+) x w+(-) crosses were transferred to 75 ml liquid medium in a 500 ml flask. They were heat-shocked for 2 hours at 50 C, then left at 37 C overnight. White ascospores from w1-(+) x w1-(-) crosses just incubated overnight at 37 C in 75 ml distilled water. At least 2000 spores were used per flask. 125 ml water were added to the flasks of germinated spores to cover the macerator probe blades fully, before maceration at full speed in a blender for 60 secs. Samples were examined with a microscope to check whether most fragments were 1-3 cells long, with further maceration if necessary. The suspension was filtered through a 305 um pore filter to remove most large fragments; some smaller fragments were also lost by trapping. The fragments in the filtrate were collected on a 0.8 um pore membrane filter, then were washed thoroughly with LMM to remove unwanted nutrients, before resuspending the fragments in 150 ml LMM in 500 ml flasks. The flasks were put on shakers and incubated for about 7 hours before filtering through a 305 æm filter to remove prototrophs, followed by a further cycle of 7 hours incubation and a final 305 um filtration. Microscopic examination was used at each stage, to monitor fragment lengths and concentrations, and reconstruction experiments using a known lysine-requiring mutation provided an additional check on the methods. 305 um was the best pore size for removing fragments larger than 1-3 cells, though a few larger fragments sometimes penetrated, end-on. The losses of different sized fragments varied with conditions, especially whether suction was applied during filtering. Typical figures for losses (as percentages of fragments of stated cell number trapped by the filter) at one 305 æm filtration were: 1-3 celled fragments, 71% loss; >3 cells, unbranched, 98%; >3 cells, branched, 100%. Monitoring of fragments showed gains in numbers of small fragments from mechanical fragmentation of hyphae during shaking to resuspend fragments after collection, and considerable losses of fragments due to fusion (or firm intertwining) of fragments, in spite of shaking during the two growth periods.

In Neurospora, Strauss (1958 J. Gen. Microbiol. 18:658-669) found that many auxotrophs died after 2 days in LMM, but our reconstruction experiments with a mixture of wild-type and lysine-requiring fragments showed almost 100% survival of both types of fragment if left unfiltered in LMM for 6 days. They also show the expected rise in percentage of auxotrophs after the first filtration. The absolute number of (non-growing) auxotrophs per flask only declined slightly after the first filtration in the 2000 um pores experiment (the 5 day result is odd) but declined much more on filtering through 305 um pores, consistent with some loss of auxotrophs by trapping in the smaller pores. The persistence of and increase in numbers of prototrophs, even after several rounds of filtration, shows fragmentation of grown prototrophic colonies during filtering, some fragments being small enough to pass through. This increase in number of prototrophic colonies did not occur in the unfiltered control flasks. The 24 hour period between filtrations in the reconstruction experiment was poorer for auxotroph isolation than the shorter periods used in other experiments.

Of six methods tried for isolating the auxotrophs after the second filtration in our main experiments, the two giving best yields were those involving a further selection stage on solid minimal medium (SMM). The best method was to spread 1 ml filtrate on a plate of SMM, then after 24 hours all small colonies, but not big ones, were isolated under a dissecting microscope into individual 75 mm tubes of solid complete medium. This gave the highest proportion of isolates with slow growth on SMM (40%) and the highest proportion of auxotrophs (about 30% of total fragments isolated). Auxotrophs included requirements for: uracil, riboflavin, inositol, aspartic acid, lysine, phenylalanine, methionine, cysteine, tryptophan, leucine, threonine, and a number of double, complex or unidentified requirements. These types are typical of previous ones obtained non-selectively in Ascobolus by Yu-Sun (1964 Genetics 50:987-997). The method also yields morphological mutants, especially ones with restricted growth. They included: dense, compact colonies, profusely branched mycelium, clock, wave, slow growth, sparse growth.

TECHNIQUES FOR COLONIAL GROWTH AND PROTOPLAST PRODUCTION IN *HUMICOLA GRISEA* VAR. *THERMOIDEA*

Humicola grisea is a thermophilic, cellulolytic fungus, with significant biotechnological potential for protein secretion. For convenient growth on solid medium, 70% reduction in colony diameter was achieved by the addition of 20 mM sodium citrate to modified Aspergillus complete medium (Pontecorvo et al. 1953, Adv. Genet. 5:141-238).

Spheroplasts were produced from mycelium by enzyme digestion with a 1:1 mixture of Novozyme 234 and cellulase CP. Spheroplast production and regeneration were studied using Mg2SO4, KCl, NaCl and sucrose and sorbitol as osmotica, but only the first of these gave detectable regeneration, and that at circa 1%. Because of this low regeneration, optimal conditions were determined for mycelial, growth time for mycelium, and enzyme concentration and time for digestion in order to produce on adequate number of viable conidia for potential transformation. This optimal protocol is given below.

Conidia were harvested from colonies grown for 7 days at 40 C by suspending in 0.85% NaCl, and inoculated at 10(5) conidia/ml into Aspergillus liquid minimal medium plus 0.5% yeast extract and 0.2% casein hydrolysate, incubated at 41 C and 150 rpm for 20.5 h. The mycelium was harvested by filtration and washed with 0.5 M Mg2SO4 in 0.02 M pH 5.6 phosphate buffer. To 125 mg wet weight of mycelium, 10.5 mg of enzyme mixture in 2.5 ml 0.5 M Mg2SO4 was added. This was incubated at 120 rpm for 180 min, then centrifuged at 500 rpm for 30 sec to remove mycelium. The supernatant containing spheroplasts was removed and centrifuged at 4000 rpm for 10 min. The spheroplast pellet was resuspended in buffered 0.5 M Mg2SO4 and recentrifuged at 4000 rpm for 10 min, repeated twice. Spheroplasts were resuspended in 1.0 ml of buffered 0.5 M Mg2SO4.

The spheroplast preparation was diluted in buffered Mg2SO4 and plated on solid Aspergillus complete medium supplemented with either sodium citrate or Mg2SO4 osmoticum, the latter adequately restricting colony size without addition of sodium citrate. Approximately 1% of spheroplasts regenerated, with more than 99% of them osmotically sensitive. This typically gave circa 10(6) regenerable spheroplasts from circa 10(8) total spheroplasts obtained from 125 mg wet weight of mycelium.

ROLE OF THE CELL WALL ON THE EXPRESSION OF OSMOTIC-SENSITIVE (*OS-1*) AND TEMPERATURE-SENSITIVE (*COT-1*) PHENOTYPES OF *N. CRASSA*. A COMPARATIVE STUDY ON MYCELIAL AND WALL-LESS PHENOTYPES OF THE "SLIME" VARIANT

Ascospore segregants ("slime"-like) of the triple mutant *fz*(fuzzy);*sg*(spontaneous germination) *os-1*(osmotic) ("slime"; Emerson 1963. Genetica 34:162-182) of *Neurospora crassa* germinate as a plasmodium which, after some time, results in a morphologically abnormal mycelium. If the mycelium of a "slime"-like isolate is cultured under high osmotic pressure (Nelson et al. 1975. Neurospora Newsl. 22:15-16), it releases cells lacking walls which proliferate as spheroplasts. Comparative studies on the production of carbon-catabolic exoenzymes in mycelial and plasmodioid phenotypes of "slime" suggest that the cell wall may have a role in the transduction of environmental signals which affect the synthesis and secretion of exoenzymes (for a review see Terenzi et al. 1990. Fungal Genetics Newsl. 37:47-49).

It is also known that cell wall synthesis inhibitors such as sorbose or polyoxin B (Selitrennikoff and Zucker 1982. Exper. Mycol. 6:65-70) inhibit the growth of wild type mycelium of *N. crassa,* but not that of "slime" spheroplasts. These results suggest that cell wall assembly may be a growth- limiting process in mycelial strains but not in wall-less ones. Along this line, we now show evidence suggesting that cell wall assembly may be involved in the phenotypic effect of mutations which cause growth inhibition either under high osmotic pressure (*os-1*) or high temperature (*cot-1*). The tester strain RCP-34 (A;*fz;sg;os-1 al-1;cot-1;nic-3*) was obtained from a cross of a "slime"-containing heterokaryon, FGSC 2713 (*A;fz;sg;os-1 arg-1 al-1 cr-1*) + (*a;tol-1 pan-1*) and BAT 9-5 (*a;cot-1; nic-3*). The mycelial phenotype (mycelial intermediate) and the plasmodioid phenotype (stable "slime") of segregant RCP-34 were obtained as described by Pietro et al. (1990. J. Gen. Microbiol. 136:121-129). These strains were cultured in liquid Vogel's complete medium, with shaking, at different concentrations of NaCl or sorbitol to test the expression of the *os-1* phenotype, or either at 25 C or 35 C to test the expression of the *cot-1* phenotype. Strains FGSC 424 (wild type), FGSC 810 (*os-1*) and BAT 9-5 (*nic-3;cot-1*) served as controls.

Figure 1A shows that increasing concentrations of NaCl inhibited the growth of the two phenotypes of RCP-34 and the *os-1* control as well. On the other hand, increasing sorbitol concentrations (Figure 1B) inhibited the mycelial intermediate of RCP-34 and the *os-1* control, but not the stable "slime" derived from RCP-34 mycelium. This result confirmed previous observations (Pietro et al. 1990). A possible explanation for the different effects of NaCl and sorbitol on the stable "slime" RCP-34 is that NaCl, but not sorbitol, besides producing an osmotic stress, produces also an ionic stress to which "slime" spheroplasts may be sensitive. After an initial lag period of approximately 16 hours, the growth rate of the mutant at 35 C diminished sharply. The same was true for the mycelial intermediate of strain RCP-34 (Figure 2B) and, in this case, the inhibitory effect temperature was rather striking, as it also was for colonies formed at 35 C on solid medium, which were much smaller than those of the *cot-1* control (not shown). In contrast, the phenotypic character of the *cot-1* mutation (restricted growth at 35 C) was not observed for the stable "slime" spheroplasts derived from isolate RCP-34, which proliferated actively at this temperature.

The primary biochemical defect of either *os-1* or *cot-1* has not yet been characterized. It has been suggested that the *os-1+* gene may play a role in cell wall assembly (Selitrennikoff et al. 1981. Exper. Mycol. 5:155-160) and this suggestion may apply for the *cot-1+* gene as well. A possible explanation for the inhibitory effects of osmolarity or temperature, respectively, on the growth of *os-1* and *cot-1* mutants is that, under restrictive conditions, a defective rate of cell wall formation becomes limiting for the rate of growth. The absence of effects of *os-1* and *cot-1* mutations in stable "slimes" reinforces this idea, and support a suggestion made by several authors (e.g. Kang and Cabib 1986. Proc. Natl. Acad. Sci. USA 83:5808-5812) that in fungal cells, cell wall assembly and cell growth are functionally coupled. Supported by FAPESP, FINEP, CNPq and CAPES.

THE BLI REGULON - A NETWORK OF BLUE LIGHT INDUCIBLE GENES OF *N. CRASSA*

Several physiological responses of *N. crassa* are observed when this fungus is exposed to blue light. Here, we do not intend to make a comprehensive list of all the light effects observed so far in *N. crassa* (for a review, see Degli Innocenti and Russo 1984. In "Blue Light Effects in Biological Systems" ed. H. Senger, Springer-Verlag. pp 213-219.), but point out only the underlying themes. First, the time interval between the light stimulus and the observed response can be very different, and ranges from a few minutes to several hours - or even days- depending on the nature of the physiological response in question. For instance, alterations in the membrane potential and the synthesis of carotenoids take place within a matter of a few minutes, whereas an increase in the number of protoperithecia in an agar plate culture, grown in conditions which favor protoperithecia production, takes 24 hours! Perithecia, the bodies containing products of the sexual cycle, if exposed to light during early stages of their formation, display phototropism of the perithecial beaks, a process which takes as many as 12 days.

A second important point to note besides the time element is that the metabolic stages of the *N. crassa* life cycle at which the tissues are exposed to blue light, and even the types of tissues exposed, are different in the physiological responses mentioned above. Thus, it appears that *N. crassa* has a genetic apparatus which can be activated by light virtually at any stage during its life cycle. We shall refer to this genetic apparatus as the bli global regulon, a conceptual equivalent of the global regulons which have been described in prokaryotic systems (Hoops and McClure 1987. In "*Escherichia coli* and *Salmonella typhimurium*" ed. F.C. Neidhardt et al. ASM Publications pp 1231-1240). The term "bli" implies "blue light inducible" to indicate the fact that in cases where the action spectrum was analyzed, only the blue part of light was observed to be effective.

Lastly, the third important point is that the bli regulon consists of a very large number of genes. A variety of translatable mRNA species appear in a temporal fashion only in the mycelia irradiated with blue light. It has been calculated that at least 60 genes are induced within 30 minutes in mycelia exposed to blue light. This massive gene action is blocked by the wc-1 or wc-2 mutations (Chambers et al. 1985. EMBO J. 4:3649-3653, Nawrath and Russo 1990. J. Photochem. Photobiol. 4:261-271). The cDNA clones of four blue light inducible (bli) genes bli-3, bli-4, bli-7, and bli-13 whose functions are not yet known, have been obtained. These bli genes define three different classes of genes based on the time taken for the accumulation of the transcripts at the "steady state level". bli-3 and bli-7 have been shown to have different DNA sequences (Eberle et al., in preparation). The lag periods observed were 2, 15 and 45 minutes respectively (Sommer et al. 1989. Nucleic Acids Res. 17:5713-5723).

The groups of genes which have been identified in other ways but are known to be regulated in response to light include al-1 (Schmidhauser et al. 1990 Mol. Cell. Biol. 10:5064-5070), al-2 (Lauter, Schmidhauser, Yanofsky and Russo, unpublished data) and al-3 (Nelson et al. 1989. Mol. Cell. Biol. 9:1271-1276) - the genes involved in carotenoid biosynthesis and con-5 and con-10 (Lauter and Russo, unpublished data) - the genes induced during conidiation (Berlin and Yanofsky 1985. Mol. Cell. Biol. 5:849-855).

How are all these light inducible genes organized in the genome and regulated precisely, in a temporal fashion? Some of the genes, including al-1, al-2, bli-3, bli-4, con-5 and con-10 are transcribed in concert in less than 5 minutes. The wc-1 and wc-2 genes which map on the linkage groups VII and I respectively, constitute an important part of the bli regulon and are believed to be the regulatory genes since none of the blue light inducible responses analyzed so far are observed in the white collar mutants. Action of the wc gene products in controlling the light inducible genes must be at the transcript level because the increase in the amount of al-1 (Schmidhauser et al. 1990. Mol. Cell. Biol. 10:5064-5070), al-2 (Lauter, Schmidhauser, Yanofsky and Russo, unpublished data), and the bli-3, bli-4, bli- 7 and bli-13 (Sommer et al. 1989. Nucleic Acids Res. 17:5713-5723) transcripts after exposure to light are not observed in the wc mutants. One possible function involved in this regulation could be protein dephosphorylation, since the wc mutants clearly show altered potentials to dephosphorylate proteins. This is the only phenotypic criterion which separates the wc-1 and wc-2 mutants (Lauter and Russo, 1990. J. Photochem. Photobiol. 5:95-103).

To date, genomic locations of eight light inducible genes have been determined in *N. crassa.* Out of three albino genes, al-1 and al-2 map on linkage group I while al-3 maps on linkage group V. Of the three con genes, con-8 maps on linkage group I, while con-5 and con-10 map on linkage group IV. We have determined the map locations of three more bli genes, bli-3, bli-4 and bli-7. Essentially, the procedure of Restriction Fragment Length Polymorphism mapping developed by Metzenberg et al. (1984. FGN 31:35-39) was followed using the strains described therein. The data will be forwarded to R.L. Metzenberg. The genomic DNA clones of the bli genes constructed in pUC vector (J□rgen Eberle, Ph.D. thesis 1990, Free University, Berlin) were used as probes. The genes bli-4 and bli-7 map at the same position, linked to ccg on LG II while bli-3 maps near con-5 and con-10 on LG IV.

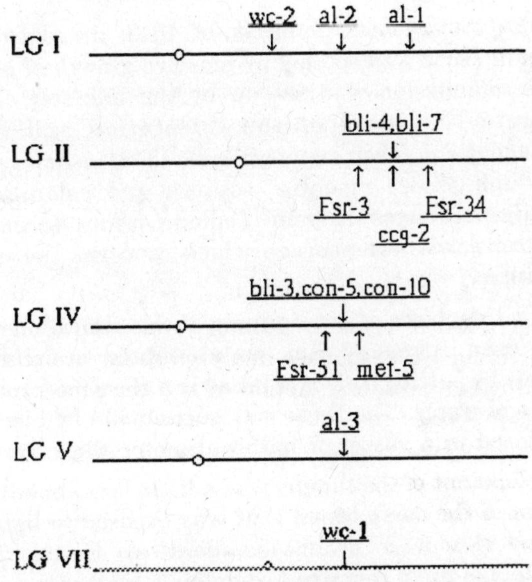

Figure. Schematic representation of the linkage groups of N. crassa. Only the relevant linkage groups are shown. Drawn above the lines are the genes belonging to the bli regulon. For the genes which have been located using the RFLP mapping technique, flanking genetic markers are shown below the line for comparison.

To summarize, we suggest that the light inducible genes of N. crassa define a global regulon, termed here as the bli regulon, which is somehow controlled at the transcript level by the wc-1 and wc-2 gene products. In addition, we have determined the map locations of three newly identified blue-light inducible genes, and find that most of the light regulated genes are dispersed fairly randomly on the genome of N. crassa. However, possible existence of small clusters containing bli-4 and bli-7 on linkage group II, and another containing bli-3, con-5 and con-10 on linkage group IV cannot be ruled out because of the limitations imposed by the available mapping techniques. It is important to note that the genes which are separated by as many as 50 kb may not always be seen as separate genes by the RFLP mapping technique.

The first published scientific study of Neurospora, including a description of photoinduction of carotenoids

In the warm, moist summer of 1842, bread from army bakeries in Paris was spoiled by massive growth of an orange mold. A commission was set up by the minister of war to investigate the cause of the infestation and to make recommendations. Their report (Payen 1843) includes a colored plate which shows mycelia, conidia and colonies of the "Champignons rouges du pain" (called *Oidium aurantiacum*). I have translated one passage which concerns the effects of illumination:

"With the object of determining if the coloration was due to light, even extremely dim, we attempted to exclude light completely by putting a piece of bread in a glass flask containing 10 grams of water. The flask was surrounded by black paper and enclosed in a vessel of half-centimeter thick bronze.

Development of the fungus was a little less abundant than on a piece of the same bread that was exposed to light under conditions that were otherwise identical. Under the first conditions, the fungi remained completely white for more than eight days (see figure b),. whereas the illuminated fungi, figure a, a', were covered with red spores. But, remarkably, the white fungi became colored when they were exposed to light for two hours."

The 1843 report names Lévillé, Montagne and Decaisne as scientists concerned with identifying the organism, while de Mirbel and one of the scientific members of the commission (Dumas, Pelouze or Payen) were concerned with microscopic and chemical analysis. Montagne (1843) independently published a drawing of the same orange fungus with a Latin description under the name *Penicillium sitophilum*.

Thermal tolerance of Neurospora was also studied at about this time (see Payen 1858, 1859). The orange spores survived 100 C for one hour and exposure to 120 C for an unspecified period. They were killed, however, at 140 C. These results were cited by Louis Pasteur (1861) in a paper reporting that mold spores could survive dry heat at 120 C to 125 C for at least an hour whereas viability was lost in a few minutes if the spores were heated in water at 100 C. These observations were relevant to the controversy then current regarding spontaneous generation.

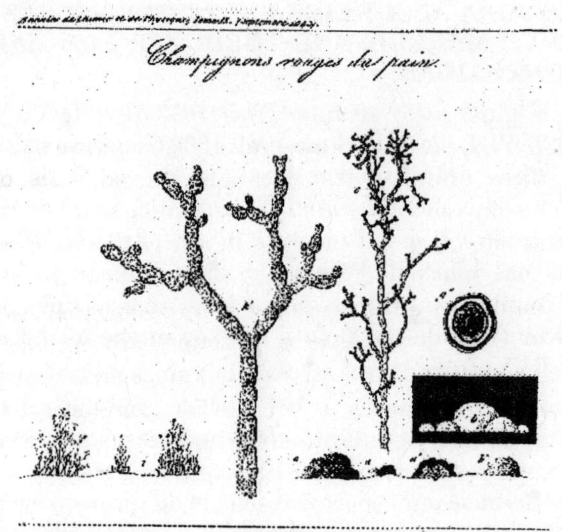

Excerpt from Plate 1 of Payen (1843). The original plate is in color. a. Colonies of the red- orange fungus *Oidium aurantiacum* as they appear to the naked eye in the cavities of infected bread. a'. A similar colony cut in two, showing in the red area a thick layer composed of innumerable small spores formed at the end of radiating filaments. The latter are yellowish white. b. Similar colonies that have grown up completely in the dark, with the result that the red color has not developed. b'. One of the colonies in b seen after exposure to light for one hour. Color begins to appear and then pigmentation progresses rapidly. c. Branching filament, about 150 x. g". Spore treated successively, under the microscope, with a dilute solution of potassium hydroxide, and aqueous alcoholic solution of iodine, then with gradually more concentrated solutions of sulfuric acis. This acid, which separates parts of the cellulose envelope that contains less nitrogenous substance, results in a blue color turning to purple, which is characteristic of the state intermediate between cellulose and dextrin. i. Normal vegetative growth as seen with the naked eye; well developed, especially under conditions of high humidity. k'. Termini of well developed filaments, showing spores and young cells.

NEUROSPORA *ALCOY* LINKAGE TESTER STOCKS WITH GROUP VII MARKED, AND THEIR USE FOR MAPPING TRANSLOCATIONS.

The original *alcoy* strains (*T(I;II)4637 al-1; T(IV;V)R2355, cot-1; T(III;VI)1, ylo-1.* Perkins et al. 1969 Genetica 40:247-278) contain three unlinked translocations tagged with markers that can be conveniently scored by inspection, without transfer. Linkage group VII is not involved in any of the translocations and was not marked. Failure to show linkage to an alcoy marker implied that the unknown was either in linkage group VII or was far removed from a marker in one of the marked groups, I-VI.

To increase efficiency, a VII marker, conidial separation-2 (*csp-2*), was introduced into *alcoy* by Perkins and Björkman (1979 Neurospora Newsl. 26:9-10). *csp-2* is scored by a simple tap-test. Because our report was part of an inconspicuous note, some labs have apparently remained unaware that the new version of alcoy is available and have continued using the less efficient originals.

We have found *alcoy;csp-2* useful not only for locating point mutants but also for determining the linkages of translocations with distal breakpoints. Before any mapping is done of a newly identified translocation we routinely determine patterns of defective ascospore in shot, unordered asci. This indicates the type of translocation, whether reciprocal or duplication-producing (Perkins 1974 Genetics 77:459-489; Perkins and Barry 1977, Adv. Genet. 19:133-285). It also tells us whether breakpoints are near centromeres or are far removed. Translocations with breakpoints closely linked to centromeres are mapped by crossing with a multicent tester (Perkins 1990 Fungal Genet. Newsl. 36:31-32). Translocations with one or both breakpoints not close to centromeres are crossed to *alcoy;csp-2,* and progeny are examined for linkage between two of the *alcoy* markers. When *alcoy;csp-2* is used, three of the 21 possible breakpoint combinations will go undetected because they coincide with the three *alcoy* translocations. Breakpoints in chromosome arms opposite those marked in *alcoy* may also not be detected. Nevertheless, the success rate has been good in a small sample of translocations tested with *alcoy;csp-2.*

Except for VII, each *alcoy* marker tags two linkage groups. When linkage is shown to one of these *alcoy* markers, a follow-up cross is required to determine which of the alternatives is correct. For mapping a point mutant, normal-sequence follow-up testers are used that have markers in the two groups in question. With an unmapped translocation, if linkage is shown between two alcoy markers in the cross by *alcoy;csp-2,* the normal- sequence follow-up testers must have four linkage groups marked unless one of the markers is *csp-2.*

A complete set of normal-sequence follow-up testers is available from the Fungal Genetics Stock Center, listed in section VII A 1 of the catalog. Triply marked follow-up testers incorporating *csp-2* are recent additions. The latter are suitable either for translocations or for point mutants that show *csp-2* linkage in crosses to *alcoy;csp-2.* With point mutants, one marker in the triply marked tester will be superfluous and can simply be ignored.

TRANSFER OF GENES AND TRANSLOCATIONS FROM *NEUROSPORA CRASSA* TO *N. TETRASPERMA*

It is difficult to obtain progeny when *N. crassa* and *N. tetrasperma* are intercrossed directly. Metzenberg and Ahlgren (1969 Neurospora Newsl. 15:9-10; 1973 Can. J. Genet. Cytol. 15:571-576) developed a transfer kit of interspecific hybrids which they used for bridging-crosses that enabled them to move the mating type genes from *N. tetrasperma* into *N. crassa.* I have recently been concerned with introgressing mutant genes and translocations from *N. crassa* into *N. tetrasperma,* and have found their strain C4,T4 a (FGSC 1778) extremely useful for the initial cross, and more fertile than other members of the kit.

A single large cross-tube usually produces enough ascospores to provide progeny for initiating a series of recurrent backcrosses to *N. tetrasperma* wild types (85 A or a; FGSC 1270, 1271) or to *N. tetrasperma* strains containing the Eight-spore gene (*E A* or *E a*; FGSC 5897, 5901).

Most ascospores are homokaryotic in the initial cross. Homokaryotic progeny are obtained in successive generations by isolating small ascospores. A majority of ascospores are homokaryotic in crosses heterozygous for *E*. Small, homokaryotic ascospores can also be obtained as infrequent exceptions from crosses with wild type *N. tetrasperma*, where they occur even in the absence of *E*.

CO-REGULATION OF TWO TANDEM GENES BY ONE BLUE-LIGHT ELEMENT IN *NEUROSPORA CRASSA*

Many genes of *Neurospora crassa* are regulated by blue light: *al-1* (Schmidhauser et al. 1990 Mol. Cell. Biol. 10:5064-5070), *al-2* (Lauter, Schmidhauser, Yanofsky, Russo unpublished), *al-3* (Nelson et al. 1989 Mol. Cell. Biol. 9:1271-1276), *bli-3, bli-4, bli-7, bli-13* (Sommer et al. 1989 NAR 17:5713-5723). For none of these genes are the blue light cis-regulatory sequences (blue-light elements, BE) known. Here we report the presence of such BE in front of *bli-4*.

The blue-light element of *bli-4* is in a stretch of DNA which is 300 bp long. We cloned this BE in front of a hygromycin reporter gene (Hygr) (Staben et al. Fungal Genetics Newsletter 36:79-81). Downstream of the reporter gene we cloned the selectable marker *tub-2* (also called *Bmlr*), which gives resistance to benomyl (Orbach et al. 1986 Molec. Cell. Biol. 6:2452-2461). The distance between the BE and the translation start (ATG) of *Bmlr* is about 3,000 bp (Fig. 1A).

After transformation in *N. crassa* wt (St. Lawrence a) we selected several benomyl resistant transformants (H1 to H4). These transformants were grown in Vogel's liquid medium with the addition of 0.5 ug/ml benomyl and total RNA of dark grown and illuminated mycelia were extracted (T. Sommer et al. 1989 NAR 17:5713-5723). The Northern blots show that both the hygromycin gene (Fig. 2b) and the *Bmlr* gene were strongly light regulated.

Experiments with a similar construct, where a tagged *bli-4* gene served as expression vector (Fig. 1b) gave different results. In the three B1-B3 transformants the tagged *bli-4* gene was strongly light regulated (Fig. 2e) while the *Bmlr* gene was only weakly regulated (Fig. 2f).

The simplest interpretation of this result is that the blue light element of *bli-4* acts as an enhancer that functions well at a distance of 3,000 bp but not as well at a distance of 4,500 bp from the translation start of the *Bmlr* gene.

A more complete understanding of blue-light regulation of gene expression in *N. crassa* will come from cloning and comparison of the properties of the blue-light element of several genes (see also Pandit and Russo 1991 Fungal Genetics Newsletter this issue).

Acknowledgements: We thank Chuck Staben and Charles Yanofsky (Stanford University) for providing the hygromycin reporter gene. We thank Uta Marchfelder for typing. This work was partially supported by the Deutsche Forschungsgemeinschaft.

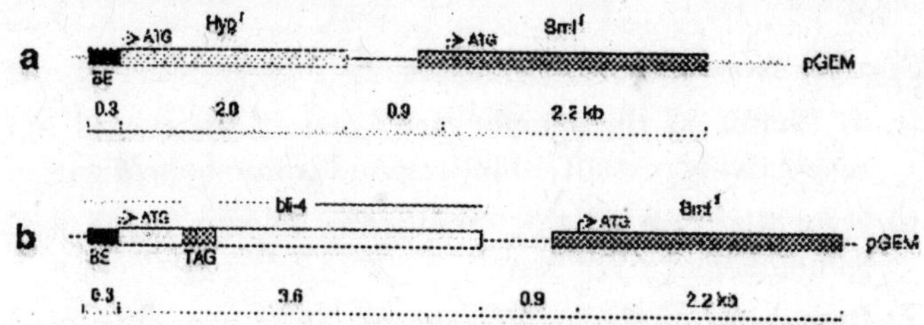

Figure.

A. Construct: blue-light box of *bli-4 Hygr* gene - *Bmlr* gene in a pGEM vector B. Construct: blue-light box of *bli-4*-tagged *bli-4* - *Bmlr* gene, in a pGEM vector

BE: Blue-light element of *bli-4* gene

Hygr: hygromycin resistant gene (*hph*)

Bmlr: benomyl resistant gene (*tub-2*)

ATG: translation start

TAG: stretch of DNA not present in *N. crassa*

pGEM: plasmid of Promega

kb: kilobase

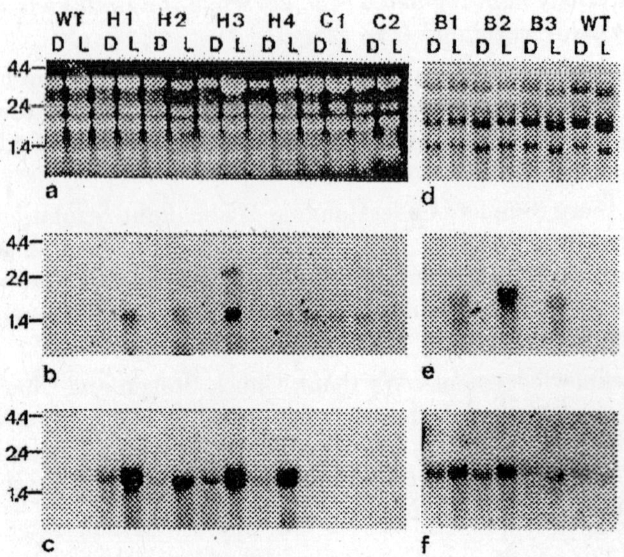

Figure. Northern Blot Analysis

a), d) Photos of the membranes used in b), c) and e),f) respectively just after blotting and before hybridization.

b) Hybridization of the membrane shown in a) with radiolabelled *hygr* gene.

c) Reprobing of the membrane a), after stripping, with radiolabelled *Bmlr* gene.

e) Hybridization of the membrane shown in d) with radiolabelled TAG DNA.

f) Reprobing of the membrane d), after stripping, with radiolabelled *Bmlr* gene.

wt: wild type (St. Lawrence, mt a)

H1-H4: four different transformants of wt with the construct a) of Fig. 1.

B1-B3: three different transformants of wt with the construct b) of Fig. 1.

C1-C2: transformants of wt with the hygromycin gene which has the Aspergillus *trpC* promoter(non-light regulated)

NITROGEN AND CARBON STARVATION REGULATE CONIDIA AND PROTOPERITHECIA FORMATION IN *NEUROSPORA CRASSA* GROWN ON SOLID MEDIA

It is known that nitrogen and carbon limitation regulate conidiation in shaken liquid cultures (Guignard et al. 1984 Can. J. Microbiol. *30*:1210-1215; Muller and Russo 1989 Fungal Genetics Newsletter *36*:58-60), and that nitrogen starvation regulates protoperithecial formation in solid medium (Sommer *et al.* 1987 Planta 170:205-208). We investigated the influence of nitrogen and carbon starvation on both sexual and asexual differentiation under exactly the same conditions of growth on solid medium.

We grew *Neurospora crassa* (OR wild type, mt *a*) on 20 ml agar medium in 8 cm diameter petri plates covered with circular dialysis membrane. The medium was a modified Vogel's medium (Russo 1988 J. Photochem. Photobiol. *2*:59-65) with 4 mM NH*4*Cl as sole nitrogen source and 1% (w/v) sorbose plus 0.1% (w/v) glucose as sole carbon sources. The plates were inoculated with 1 x 10 *4* conidia spread onto the dialysis membranes and were incubated in the dark at 23*o*C. After three days the membranes were transferred onto new plates. The four different types of plates were:

+N +G (50 mM NH*4*Cl, 1% sorbose, 0.1% glucose)

+N -G (50 mM NH*4*Cl, no carbon source)

-N +G (no nitrogen source, 1% sorbose, 0.1% glucose)

-N -G (no nitrogen source, no carbon source)

All plates were illuminated with 1 min of blue light (6 W/m *2*) just after transfer and 24 h later. Protoperithecia were counted using a binocular microscope 48 h after the transfer (Degli Innocenti and Russo 1983 Photochem. Photobiol. 37:49-51) and macroconidia were collected with 5 ml sterile twice-distilled water. The number of viable macroconidia was determined immediately by plating. The data are shown in Table 1 and analyzed in Table 2.

Nitrogen starvation inhibited production of conidia and induced production of protoperithecia. Carbon starvation induced both conidia and protoperithecia. In the case -N(-G/+G) the ratio was 1, but this is probably due to saturation of the biological system. It was not possible for cultures to make more than 10,000 protoperithecia/plate and this value is reached already by nitrogen starvation.

Nitrogen and glucose acted synergistically for conidial production. For example, the plates of +N-G type had 1 x 10 *4* more conidia than the plates of -N+G type. It would be interesting to know whether glucose and nitrogen starvation influenced the expression of con and bli genes. *con* genes are candidates as important genes in the conidiation process (Berlin and Yanofsky 1985 Mol. Cell. Biol. 5:849-855) while bli genes are candidates as important genes in the formation of protoperithecia (Sommer et al. 1989 NAR 17:5713-5723).

5

A New Auxotrophic Mutant

A mutant auxotroph for methionine was isolated in *Podospora anserina* during a transformation experiment. The transforming plasmid (pPAaURA5) consisted of a 1.55 kb nuclear DNA fragment of *Podospora* containing the *URA5* gene (Begueret et al. 1984, Gene 32:487-492; Turq and Begueret 1987, Gene 53:201-209), and most (2.06 kb) of the Podospora intronic alpha mitochondrial sequence (Osiewacz and Esser 1984, Curr. Genet. 8:299-305) cloned in the pUC18 vector. The recipient strain carried the mutant *ura5-6* allele (Razanamparany and Begueret 1986, Curr. Genet. 10:811-817) and the mating plus locus (*mat+*). [In *Podospora*, transformation with plasmids occurs by integration of the vector mainly outside the resident locus (Brygoo and Debuchy 1985, Mol. Gen. Genet. 200:128-131; Razanamparany and Begueret 1986, Curr. Genet. 10:811-817).]

After transformation of the *ura5-6 mat+* strain with plasmid pPAaURA5 and selection of primary (*ura+*) transformants, the transformants were crossed to a *ura5-6 mat-* strain in order to purify the transformant nuclei through meiosis. In most cases, 50% *ura+* and 50% *ura-* spores were obtained in the progeny. However, one primary transformant gave different results: of 46 monocaryotic spores tested, 20 were auxotrophic for uracil, the 26 others were methionine auxotrophs.

Three purified *ura+ met-* transformants were crossed with wild-type. The *met-* phenotype segregated as a single recessive gene. The percentage of second division segregation was about 60%. *ura-* spores were obtained in the progeny indicating that the parental transformant strain contained the *ura5-6* allele.

Furthermore, the presence of tetrads containing 2 dicaryotic spores (*ura- met+*) and 2 dicaryotic spores (*ura+ met-*) indicated the integration event of the URA5 gene occurred in a chromosome different from that carrying the URA5 locus. Three purified transformants (*ura+ met-*) were crossed with the *ura5-6* strain. In the progeny, the two phenotypes (*ura+* and *met-*) were associated and segregated together.

These results indicate that the *met-* phenotype of the primary and purified transformants resulted from the integration of the URA5 gene of plasmid pPAaURA5 in a gene involved in methionine biosynthesis. Analysis of the genomic DNA of this transformant has not been completed; we do not know if the entire plasmid or only the URA5 gene has been integrated. The strain has been called *met1*. It is quite fertile and stable through vegetative growth as well as through meiosis. It grows on minimal medium supplemented with methionine, cysteine or homocysteine, but is not complemented by O-acetyl-homoserine. It constitutes the first example in *Podospora* of the isolation of mutants by transformation-mediated gene disruption.

LIPOFECTIN INCREASES THE EFFICIENCY OF DNA-MEDIATED TRANSFORMATION OF *NEUROSPORA CRASSA*

Transformation of *N. crassa* using the calcium chloride/polyethylene glycol procedure (Fincham 1989. Microbiol. Rev. 53:148-170) can yield 100-10,000 transformants per microgram of input DNA when integration occurs nonhomologously, and considerably lower numbers of transformants when homologous integration of DNA is required. Increasing the efficiency of transformation would be useful for screening cosmid libraries and for obtaining transformants from homologous integration events. In addition, strains that transform poorly using standard procedures, e.g. the cell-wall-less derivatives of *os-1* (Phelps et al. 1990 Curr. Microbiol. 21:233-242) would benefit from improvements in the transformation procedure. Lipofectin, a liposome preparation formulated from cationic lipids [Bethesda Research Laboratories, Gaithersburg, MD], has been reported to increase transformation of the yeast *Schizosaccharomyces pombe* by 35-fold with plasmids or large (>500 kb) linear DNA molecules (Allshire 1990 Proc. Natl. Acad. Sci. USA 87:4043-

4047). In an effort to increase the transformation efficiency of *N. crassa*, we tested the effects of including lipofectin in a standard transformation procedure.

Germinated macroconidia of wild type *N. crassa* (74-OR23-1VA) were grown, harvested and counted. Protoplasts were obtained by treatment with Novozym 234 as described (Orbach et al. 1986 Mol. Cell. Biol. 6:2452-2461). Protoplasts of the *os-1* derivative TM-1 were grown and harvested as described (Phelps et al. 1990 Curr. Microbiol. 21:233-242). Protoplasts of both strains were washed twice with 1 M sorbitol and once with STC (1 M sorbitol, 50 mM Tris-HCl pH 8.0, 50 mM CaC12). Protoplasts were resuspended in STC (0.4 ml per 5 x 10(7) cells). Then 0.1 ml PTC (40% polyethylene glycol 3300 [Sigma Chemical Co., St. Louis, MO], 50 mM Tris-HCl pH 8.0, 50 mM CaC12) and 5 ul of dimethyl sulfoxide were added per 5 x 10(7) cells. Competent protoplasts were stored at -80 C after aliquoting into eppendorf tubes; aliquots in tubes were transferred directly from wet ice to the -80 C freezer.

For transformation, cells were thawed on ice. DNA was premixed with heparin (25 ul; 125 ug) in 12 x 75 mm polystyrene tubes [Falcon 2058 tubes; Becton Dickinson Labware, Lincoln Park, NJ]; cells (0.1 ml) were added and the mixtures incubated on ice for 30 min. Lipofectin was added, and incubation continued for 15 min at room temperature. One ml of PTC was added, and after gentle mixing, suspensions were incubated 20 min at room temperature. Aliquots were added to molten regeneration agar overlayed on minimal plates containing 1 ug/ml benomyl (DuPont, Wilmington, DE) or 200 ug/ml Hygromycin B (Boehringer Mannheim, Indianapolis, IN). Plates were scored after 48 h incubation at 34 C (for wild type) or after 96 h at 25 C (for TM1).

Inclusion of lipofectin at a concentration of 35 ug per ml of wild type protoplasts in the standard protocol increased yields of transformants by nearly 10-fold when expression of a semi-dominant allele of the *tub-2* gene (Orbach et al. 1986 Mol. Cell. Biol. 6:2452-2461), which confers benomyl resistance, was selected. Lipofectin at concentrations higher than 50 ug/ml did not result in increases in transformation (not shown). Similar results were obtained using a library of cosmid DNA to transform *N. crassa* to hygromycin resistance. The efficiency of transformation of TM-1 protoplasts to benomyl or hygromycin resistance, using pMO63 or pMOcosX cosmid library DNA, respectively, increased 3-4 fold (average of 4 transformations) when lipofectin was included in the transformation procedure (data not shown). Addition of PTC appeared necessary for efficient transformation of *N. crassa* with or without added lipofectin; preincubation of DNA with lipofectin prior to addition of spheroplasts did not increase transformation efficiency compared to controls lacking lipofectin. We have limited results concerning the stability of recovered transformants in that 15/15 colonies recovered from primary transformation plates were stable (i.e., grew on hygromycin-containing medium for at least two transfers.

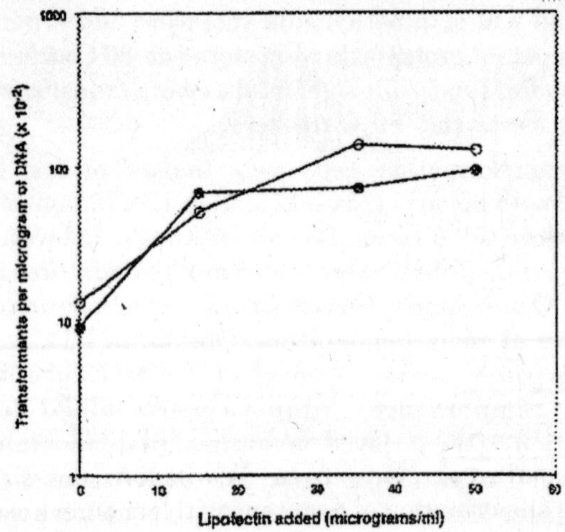

Figure: Effects of lipofectin concentration on transformation of wild type *N. crassa* with a plasmid conferring benomyl resistance. Wild type protoplasts were incubated with 2 ug (open symbols) or 5 ug (closed symbols) of pMO63 DNA and 0, 15, 35, or 50 ug/ml of lipofectin. The number of benomyl-resistant colonies obtained per microgram of DNA at each lipofectin concentration is plotted as a function of lipofectin concentration. pMO63 contains the mutant *tub-2* gene that confers benomyl resistance on a 3.1 kb *Hin*dIII fragment.

Our results show that addition of lipofectin to a standard *N. crassa* transformation protocol increases the number of recovered transformant colonies 3-10 fold. These increases should prove helpful for experiments in which the highest possible transformation efficiencies are desired.

Acknowledgements: We thank Dr. N. Raju for processing. CPS thanks Dr. David Perkins and Dr. Charles Yanofsky for their gracious hospitality while on sabbatical leave at Stanford University. MSS thanks Dr. Yanofsky for helping to support his postdoctoral study. This work was supported in part by NIH and NSF grants to CPS.

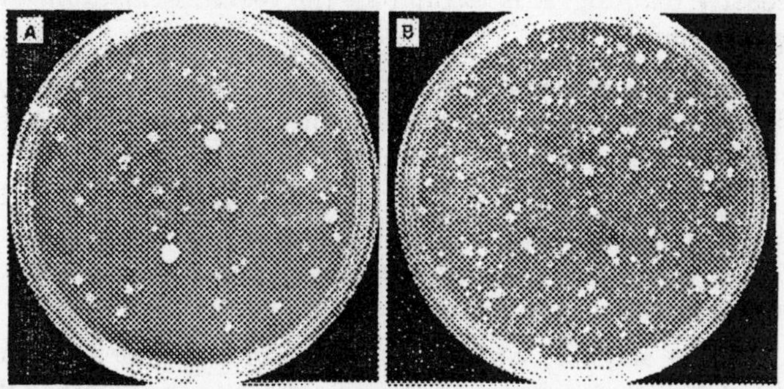

Figure: Lipofectin increases the efficiency of transformation of *N. crassa* to hygromycin resistance with cosmid DNA. Wild type protoplasts (0.1 ml) were incubated with 1 ug of pooled cosmid DNA in vector pMOcosX and (A) no lipofectin or (B) 35 ug/ml lipofectin. One-tenth of the transformation mixture was plated on each plate. The pMOcosX library is available through the FGSC.

TERGITOL-INDUCED COLONIAL GROWTH WITHOUT INHIBITION OF CONIDIATION.

The cloning of genes from *Neurospora crassa* usually involves transformation of a mutant with genomic library DNA and the subsequent identification of transformants showing wild-type phenotype (Vollmer and Yanofsky 1986. Proc. Natl. Acad. Sci. 83:4869- 4873).

This can be problematic in the case of morphological mutants, because the sorbose- containing medium that is typically used to promote colonial growth of the transformants tends to interfere with conidiation, and sometimes inhibits aerial hyphal growth altogether. The medium described below uses the nonionic detergent tergitol (Tatum et al. 1949. Science 109:509-511) to induce colonial vegetative growth without affecting aerial hyphal growth or morphology.

This medium is based on standard Vogel's minimal medium plus 1.5% sucrose and 1.5% agar. Tergitol NP-10 (available from Sigma) is added to autoclaved bottom agar medium at a concentration of 0.005%. Medium containing 0.001% tergitol allows some spreading growth, while viability suffers at concentrations of 0.01% and above. Sterilization of the concentrated tergitol has not been necessary. Care should be taken when mixing the detergent into the medium to avoid excessive foaming. It is normal for the resulting medium to be slightly turbid.

Plating transformants directly in top agar containing tergitol will kill the spheroplasts. Instead, the top agar should consist of Vogel's minimal medium plus 1.5% sucrose, 1.5% agar, and 1 M sorbitol for osmotic stabilization. Tergitol medium has successfully been used with both benomyl and hygromycin to differentiate wild-type transformants from aconidial mutant transformants.

Unlike sorbose, which restricts both mycelial and aerial hyphal growth, tergitol only restricts the growth of hyphae that are in contact with the medium. Aerial growth, which commences about one day after plating, proceeds vigorously until aerial hyphae and conidiophores completely fill the space between the agar and the lid of the Petri dish. The colonies must therefore be checked about twice a day to avoid overgrowth. For this reason, standard sorbose medium is still preferable for colonial growth where the morphology of the transformants is inconsequential. However, the preservation of aerial morphology makes tergitol medium quite useful for the study of morphological mutants.

A System for Increasing Variability in Filamentous Fungi?

The possibility of genetic analysis in the Fungi Imperfecti had its origin with the description of genetic recombination without sexual reproduction, later designated the parasexual cycle, in the ascomycete *Aspergillus nidulans* (Pontecorvo et al. 1953. Adv. Genet. 5:141-238; Pontecorvo 1956. Ann. Rev. Microbiol. 10:393-400). As the description occurred in a species with sexual reproduction, it obviously allowed comparisons between data obtained from sexual and parasexual analysis.

Therefore it was possible to establish the basic parameters of genetic analysis far more easily that would have been possible if the parasexual cycle had been discovered in an imperfect species.

Typically, the parasexual cycle is a sequence of events as follows: Heterokaryosis occurs, followed by fusion of haploid nuclei in non-specialized structures producing relatively stable diploid nuclei. The resulting diploid colonies can be isolated as prototrophs using suitable selective procedures (Roper 1952. Experientia 8:14-15). Diploid nuclei can spontaneously generate haploid recombinants by non-disjunction, and/or diploid recombinants by mitotic crossing over, both events occurring at low frequencies (Pontecorvo and Käfer 1958. Adv. Genet. 9:71-104; Käfer 1961. Genetics 46:1581-1609). These low frequencies may well be of importance in imperfect species, or in heterothallic species in which compatible mating types get together infrequently. However it almost certainly means that, from a genetic diversity point of view, the parasexual cycle is a comparatively small effect in a species which is constantly producing spores by meiosis, as in *A. nidulans* (Fincham et al. 1979. Fungal Genetics. Blackwell Sci. Publications).

Following the description of the parasexual cycle in *A. nidulans*, the same recombination mechanism was found in several other species, and although variations were detected in some species, they were interpreted in light of the classical sequence of events. For instance, in Aspergillus niger the frequency of prototrophic colonies, presumed to be diploids, was higher than the frequency reported for *A. nidulans* (Pontecorvo et al. 1953. J. Gen. Microbiol. 8:198-210). Also in *A. niger* some prototrophic colonies were considered to be stable diploids as no spontaneous segregation was detected (Chang and Terry 1973. Appl. Microbiol. 25:890-895; Das and Ilczuk 1978. Folia Microbiol. 23:362-365). Lhoas (1967. Genet. Res. 10:45-61) and Bos et al. (1988. Curr. Genet. 14:437-443) pointed out that prototrophic colonies should be analyzed as soon as possible or even discarded, otherwise anomalous segregation patterns were likely to be detected. Lhoas (op. cit.) also reported a high frequency of mitotic recombination in *A. niger* when compared to *A. nidulans*. A more extreme variation, namely a transient diploid state, was proposed in *Acremonium chrysogenum* since recombinants (possibly haploids and hyperhaploids) were isolated directly from a heterokaryotic stage which was itself transient (Ball and Hamlyn 1978. Brazil. J. Genet. 1:83-96).

Working with a citric acid-producing strain of *A. niger*, Bonatelli Jr. et al. (1983. Brazil. J. Genet. 6:399-405) detected at least three types of prototrophic colonies arising from balanced heterokaryons. These were classified as haploid recombinants, diploids heterozygous for only some of the genetic markers and typical diploids as determined by the benlate test (Upshall et al. 1986. J. Gen. Microbiol. 100:413-418), conidial diameter, DNA content per nucleus and segregation analysis. Based on these data it was suggested by Bonatelli Jr. and Azevedo (1990.

Abstracts 4th IMC, Regensburg, FRG. p. 143) that in *A. niger* besides the typical heterozygous diploid nuclei another diploid state might also occur originating from haploid and/or diploid recombinants in the heterokaryotic hyphae. These diploid nuclei are possibly committed to a process that could generate recombinants, which resembled meiosis but was not accompanied by differentiation of typical sexual structures, as is usual for sexual fungi, nor having the same complex genetic control. That process, which seems to be a variation of the parasexual cycle, was tentatively designated parameiosis (Bonatelli Jr. et al. 1983. Brazil. J. Genet. 6:399-405).

More recently, in a survey of prototrophic *A. niger* colonies from crosses involving multi-marked strains described by Masiero and Bonatelli Jr. (1989. Brazil. J. Genet. 12:707- 718) and Bos et al. (1988. Curr. Genet. 14:437-443), Calil and Bonatelli Jr. (data not published) also detected diploid colonies showing homozygosis for genetic markers. Furthermore, these workers also isolated diploid colonies which showed a frequency of mitotic crossing over two to eight fold higher than expected in two tested intervals. These data seem to corroborate the idea that different prototrophic colonics can arise from heterokaryons of that species and also that recombination can occur at relatively high levels at least in some diploid strains indicating that different types of diploid nuclei might exist.

Variations in the parasexual cycle have been described in other species besides *A. chrysogenum* and *A. niger*. In

Metarhizium anisopliae (Bergeron et al. 1982. Can. J. Genet. Cytol. 24:643-651; Silveira and Azevedo 1987. Enz. Microb. Technol. 9:149-152; Bagagli et al. 1991. Brazil. J. Genet. 14:261-271); *Magnaporthe grisea* (Crawford et al. 1986 Genetics 114:1111-1129); *Beauveria bassiana* (Paccola-Meireles and Azevedo 1991. J. Inv. Pathol. 57:172-176); *Trichoderma sp.* (Stasz and Harman 1990. Exp. Mycol. 14:145-159; Furlaneto and Pizzirani-Kleiner 1992. FEMS Microb. Letters 90:191-196) and *Fusarium oxysporum* (Molnãr et al. 1990. Mycol. Res. 94:393-398) the emergence of mainly haploid and hyperhaploid recombinants directly from heterokaryons occurs at frequencies which considerably exceed the usual values of the parasexual cycle as in *A. chrysogenum.*

That situation seems to indicate the occurrence of a high proportion of transient diploid nuclei, because very few if any diploid colonies were detected. For these cases mechanisms other than parameiosis have been proposed, e.g. a primitive meiosis (Bonatelli Jr. and Azevedo 1990. Abstracts 4th IMC, Regenburg, FRG p. 143; Bagagli et al. 1991. Brazil. J. Genet. 14:261-271) or even a mechanism of a natural intertransformation, as suggested in Trichoderma (Stasz and Harman 1990. Exp. Mycol. 14:145-159) and at this stage they cannot be conclusively ruled out. In *Fulvia fulva* (Talbot et al. 1988. Curr. Genet. 14:567- 572) and *Verticillium sp.* (Jackson and Heale 1987. J. Gen. Microbiol. 133:3537-3547) the situation is more similar to *A. niger* where haploid and/or diploid recombinants can occur in lower frequencies than in *A. chrysogenum* together with typical and non-typical diploid colonies. It seems to indicate that only some diploid nuclei in the heterokaryotic hyphae are transient and consequently committed to generate recombinants.

In all these cases, however, the existence of transient diploid nuclei can be suggested and that might be a process for obtaining variability in Fungi Imperfecti and also, as is the case of *M. grisea,* in heterothallic species where mating types occur apart from each other most of the time (Crawford et al. op cit.). It is interesting to note that in *A. nidulans*, where the parasexual cycle was first described, no evidence of a particularly unstable diploid state has been found, despite intensive genetic analysis over many years.

PRODUCTION OF MUTANTS OF *GAEUMANNOMYCES GRAMINIS* VAR. *TRITICI* AND VAR. *AVENAE* BY 4-NITROQUINOLENE-OXIDE TREATMENT OF PROTOPLASTS.

The ascomycete fungus *Gaeumannomyces graminis* is the causative agent of take-all disease of cereals. Much information about the physiology and pathology of this organism has been generated (Asher and Shipton (Eds.) "Biology and Control of Take-All", Academic Press, 1981), but genetic studies such as the production of mutants have been hindered by problems in obtaining viable propagules suitable for mutagenesis (Blanch et al. 1981. Trans. Brit. Mycol. Soc. 77:391-399). The fungus is homothallic but many strains cannot be induced to form perithecia in culture and even the fertile strains produce insufficient numbers of ascospores for use in mutagenesis. It is, however, possible to produce and regenerate large numbers of protoplasts and Rochefrette et al. (1979. Ann. Phytopath. 11:43-51) reported the mutagenesis of *G. graminis* protoplasts using nitrosoguanidine (NTG). Here we report the mutagenesis of protoplasts by 4-nitroquinolene oxide (NQO), reportedly a safer and more disposable mutagen than NTG (Bal et al. 1977. Mutat. Res. 56:153-156).

Protoplasts were prepared by standard methods (e.g. Ballance and Turner 1985. Gene 36:321-331) with the following modifications. Protoplasts were produced from washed 2-day old liquid cultures (potato dextrose broth) by shaking (100 rpm) in 0.6 M KCl, 5 mg Novozyme 234/ml for 4 h at 25 C. Mycelial debris was filtered out with four layers of Miracloth (Calbiochem Inc, USA) and the resulting suspension pelleted by centrifugation at 3000 x g. The supernatant was removed and the protoplasts were washed by resuspending them in 0.6 M KCl and repelleting them three times.

Approximately 108 protoplasts were incubated for 1 h at room temperatures in 5 ml 0.6 M KCl containing NQO (1 ug/ml Sigma) from a 10 mg/ml stock in acetone (this treatment was predicted to kill approximately 99% of protoplasts; see Figure 1). After treatment, the NQO was neutralized by the addition of an equal volume of 5% sodium thiosulphate in 0.6 M KCl, protoplasts were washed three times, as previously described, in protoplast regeneration medium (per liter: NaNO3,

4 g; KCl. 44g; KH2PO4, 1.52 g; $MgSO_4.7H_2O$, 1 g; glucose, 10 g; yeast extract, 0.5 g; casamino acids, 0.2 g; $FeSO_4.7H_2O$, 1 mg; $ZnSO_4.7H_2O$, 10 mg; $CuSO_4.5H_2O$, 1 mg; $Na_2B_4O_7.10H_2O$, 1 mg; and (NH_4)6Mo7O24.4H2O, 50 mg; adjusted to pH 6.5) and added to 5 ml molten regeneration medium containing 0.5% agar and the appropriate inhibitor, which had been cooled to 48 C. The protoplast/agar mix was then overlayed onto protoplast regeneration agar plates.

The proportion of protoplasts containing nuclei was estimated in a separate experiment by bis-benzimide staining and visualization under UV light irradiation. Approximately 16% contained one nucleus and 6% contained two or more nuclei. Regeneration frequencies for protoplasts of both *avenae* and *tritici* varieties were normally between 5 and 10%.

Plates were incubated at 22 C for 1-3 weeks and fast growing colonies were picked and retested on the appropriate selective media. Rates of transformation of protoplasts to phleomycin resistance can be enhanced by regenerating protoplasts for a few days before adding the antibiotic to the plates. In our hands, however, it was found that regeneration of mutated protoplasts before the addition of inhibitor resulted in high levels of background growth which interfered with selection of resistant colonies. The following mutants resistant to these inhibitors were selected:

1280 *G. graminis* var. *tritici* colonies surviving NQO treatment were screened for auxotrophy; 1 methionine and 1 arginine auxotroph were isolated together with a mutant requiring NH2+ as a nitrogen source. In addition, 6 color and 5 morphological mutants were isolated from the *tritici* isolate, and 2 color and one morphological mutant were isolated from the avenae isolate.

In conclusion, this method has provided us with an effective means of mutagenizing this organism: 195 mutants were produced using NQO mutagenesis compared with only 7 mutants reported for NTG mutagenesis (Rochefrette et al.) although differences in selection procedures and strains used may render this comparison invalid. No direct comparison of NTG and NQO mutagenesis rates was done. Work is in progress to characterize the mutants so far isolated.

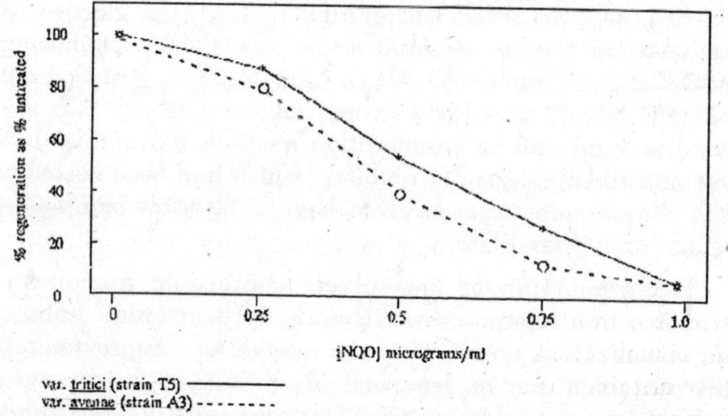

Figure. Protoplast survival after NQO treatment for 1 hour (Regeneration frequency is expressed as a percentage of untreated protoplast regeneration - typically 5-10% of untreated protoplasts regenerated)

A CONVENIENT AND SAFE METHOD TO GENERATE A LARGE QUANTITY OF *ASPERGILLUS PARASITICUS* SPORES.

In order to examine the spore pigments in strains of *Aspergillus parasiticus*, we developed a method to safely generate a large quantity of spores while conserving laboratory space, resources, and time.

Traditional methods employed to generate stock supplies of spores usually involve inoculating the surface of nutrient agar, in a flask or petri dish, with a defined number of spores (Pontecorvo et al. 1953. Adv. Gen. 5:141-238). The culture is placed at 30 C for 48 hrs to encourage mycelial growth then moved to room temperature. Asexual sporulation is evident after an additional two days with a maximum number of spores generated by ten days. The traditional vessels used for large scale spore production are 2.8 liter Fernbach flasks (surface area-SA = 285 cm2) and 15 cm petri dishes (SA = 175 cm2). Spores are harvested by placing buffer in the flask or dish and gently scraping the agar surface with a long handled glass rod. Initially the hydrophobic spores are difficult to wet and repeated mechanical agitation is required to harvest them. Finally, the spore suspension is collected, washed, and concentrated by centrifugation. During the protracted scraping step, spores can escape from the vessel thereby increasing the chances of laboratory contamination and human exposure to a concentrated spore inoculum. The method we describe eliminates the scraping step and employs easily stacked, reusable, non-breakable plastic containers with three times the surface area of a Fernbach flask. We feel that the risk of releasing a large quantity of spores while harvesting is greatly reduced.

Two liter, polystyrene Corning Roller Bottles (RBTL - catalogue number = 25240) are inoculated with 10(7) spores in 200 mls of 3% YGT agar (yeast extract 0.5%, glucose 2%, and trace elements) at 50 C using standard sterile technique. The caps are replaced, and the bottles rolled steadily in an ice bath until the agar has solidified along the inner surface (3 to 5 minutes). The caps are then replaced with sterile foam plugs to allow gas exchange and incubated as described above. The bottles may be placed in a vertical position or conveniently stacked on their sides during mycelial growth and sporulation. Normal mycelial growth initially proceeds within the agar and eventually breaks through the surface where asexual sporulation occurs. The spores are harvested by placing 200 mls of buffer inside the vessel, replacing the original plastic cap, and shaking vigorously. The spore solution is decanted, concentrated by low speed centrifugation, and washed in storage buffer. The spores are almost entirely free of contaminating mycelial debris and agar fragments.

Table 1

Spore Inoculation	Spore Yield per Culture Vessel: Fernbach	Roller Bottle	Factor Increase
10^4	$5.1 +/- 1.4 \times 10^9$	-----	
10^7	$4.7 +/- 0.2 \times 10^9$	$1.9 +/- 0.18 \times 10^{10}$	4x
3×10^7	-----	$1.65 +/- 0.61 \times 10^{10}$	
Surface inoculation in buffer			
3×10^7	-----	$2.18 +/- 0.81 \times 10^{10}$	

Table 1 indicates typical spore yields from Fernbachs and RBTLs inoculated with a variable number of spores. The initial spore inoculum did not play a statistically significant role in the final number of spores generated per flask or RBTL. The number of spores generated was roughly proportional to the available surface area.

Although inoculating pre-poured RBTLs with a spore solution may seem more convenient, the harvesting procedure is compromised in two ways. An additional filtration step is required to remove debris that is released from the mycelial mat growing on the agar surface, and bits of agar frequently plug the filters. When spores are inoculated with the molten agar, mycelial filaments infiltrate and effectively reinforce it. A surface mycelial mat leaves the agar fragile and susceptible to the shear forces required to wet the spores.

In summary, the described procedure has a number of benefits over conventional methods for generating large quantities of spores. It is safer due to the unbreakable nature of plastic RBTLs and because spore harvesting can be done in a completely closed container. In addition, the RBTLs save space since they have approximately 3X the usable surface area compared to Fernbach flasks and they can easily be stacked on their side on a shelf.

SCREENING FOR LIGNIN PEROXIDASE GENES IN NATURAL ISOLATES OF WHITE ROT FUNGI.

To find lingin peroxidase genes similar to those of *Phanerochaete chrysosporium,* we have studied several species of white rot fungi collected in nature. The methodology has been DNA hybridization techniques using two synthetic oligonucleotides with a sequence that corresponds to a fragment of the H8 lignin peroxidase gene from *P. chrysosporium* (Schalch et al. 1989. Mol. Cell. Biol. 9:2743-2747) controlling the structure of the predominant form of the enzyme in this fungus.

The white rot fungi *Pycnoporus cinnabarinus* (strain Pc58), *Pycnoporus sanguineus* (strain Ch470), *Bjerkandera adusta* (strain Ch1080), *Coriolus versicolor* (strain Ch730) and *Pleurotus sp.* (strain Ch1040) were isolated from carpophores collected in the rain forest of southern Chile, and now are kept in our laboratory collection.

In the laboratory, these strains were grown for 10 days at 30 C in liquid medium containing 1.5% malt extract, 0.75% yeast extract and 0.4% glucose, on a rotary shaker at 120 rpm. Mycelium was harvested on Whatman #1 paper and washed with distilled water, frozen at -20 C and ground in a mortar,

DNA was prepared by the rapid lithium chloride method (Leach et al. 1987. FGN 34:32-33), treated with 30 ug/ml RNAase A at 37 C, followed by three phenol:chloroform:isoamyl alcohol (25:24:1) extractions as an additional step. DNA was dialyzed against TE buffer (10 mM Tris-HCl, 1 mM EDTA) for 24 h. DNA samples, 4 ug, were digested with *Bam*HI at 37 C for 3 h. Then, two phenol: chloroform:isoamyl alcohol (25:24:1) extractions were performed, DNA samples, in 100 ul of TE buffer, were denatured, transferred to a nylon membrane (Pall), treated with UV light for 5 min and prehybridized for 10 h at 48 C. The 50-base synthetic oligonucleotides used as probes were:

I. AGCTCCAGAA GCCATTCGTT CAGAGGCACG GTGTCACCCC TGGTGACTTC

II. ATCGCCTTCG CTGGTGCTGT CGCGCTCAGC AACTGCCCTG GTGCCCCGCA

The oligonucleotides were mixed and end labeled with polynucleotide kinase and [32P]-gamma-ATP. Hybridization was carried out at 48 C for 24 h. An additional probe was pAT153 containing a cDNA for H8 lignin peroxidase labeled with biotin.

Table below shows the results of hybridization experiments using *Phanerochaete chrysosporium* and plasmid DNA as positive hybridization controls and *Neurospora crassa* DNA as a negative control, respectively.

Only two species exhibited hybridization with the probes used. These species were *Bjerkandera adusta* (Ch1080) and *Coriolus versicolor* (Ch730). In contrast, *Pycnoporus cinnabarinus* (Pc58) and *Pycnoporus sanguineus* (Ch470), which have significant lignin degrading capacities, did not show hybridization with these two oligonucleotides.

Other fungi, such as *Pleurotus sp.* (Ch1040) and *S. commune* (Ch29), gave only faint hybridization signals that were insufficient to suggest the presence of lignin peroxidase genes. Although both *Pyconoporus* and *Pleurotus* species are classified as white rot fungi, these results suggest the presence of a different lignin degrading system in these fungi. *P. cinnabarinus* and *P. sanguineus* have high levels of phenol oxidases, such as laccase and when grown in a medium containing ground wood of *Nothofagus dombeyi* plus malt extract, they utilize such substrate efficiently.

Table. Dot blot analysis of Chilean natural isolates of lignolytic fungi.

	DNA probe	
	Oligonucleotides	pAT153-H8
Phanerochaete chrysosporium BKM-1767	+++	+++
Pycnoporus cinnabarinus (Pc58)	–	–
Pycnoporus sanguineus (Ch470)	–	–
Coriolus versicolor (Ch730)	++	++
Bjerkandera adusta (Ch1080)	+	+
Schizophyllum commune (Ch29)	–	–
Pleurotus sp. (Ch1040)	–	–
Neurospora crassa 74A	–	–
pAT153-H8	+++	+++

6

A Microbiological Assay for Host-specific Fungal Polyketide Toxins

Genetic analysis of biosynthetic pathways for fungal secondary metabolites depends on availability of efficient and dependable assays for the end products. Some fungal plant pathogens produce secondary metabolites called host-specific toxins. Until recently, all bioassays for these toxins required use of whole plants or plant parts (Yoder 1981 In: Toxins in Plant Disease, Durbin ed., pp. 45-78). Since host-specific toxins, by definition, affect only plants that are susceptible to the toxin-producing fungus, other plants, animals and microorganisms are not sensitive and therefore cannot be used in bioassays. An opportunity to circumvent this problem arose when the cellular target of T-toxin (produced by *Cochliobolus heterostrophus*) and PM-toxin (produced by *Mycosphaerella zea-maydis*) was discovered to be a 13 kD polypeptide encoded by the corn mitochondrial gene *Turf13. E. coli* cells expressing *Turf13* are sensitive to these toxins (Dewey et al. 1988 Science 239:293- 295), both of which are polyketides with similar structures and biological activities. We report here the use of toxin-sensitive *E. coli* cells for both qualitative and quantitative toxin bioassays.

Cells of strain DH5alpha carrying the *Turf13* plasmid pATH13-T are recovered from storage in 25% glycerol at -70 C (cells can be used for no longer than 1 week after recovery from storage), streaked on an L plus ampicillin (100 ug/ml) plate, incubated overnight, restreaked from single colonies on L plus ampicillin, inoculated into 100 ml liquid core medium [M9 salts and per liter: 1 M MgSO4 (1 ml), 1 M CaCl2 (0.1 ml), casamino acids (5g)] plus ampicillin and tryptophan (20 mg/ liter, filter sterilized), and incubated with shaking (250 rpm) at 30 C overnight. Cells are diluted 1:10 with core medium plus ampicillin (OD550 ~ 0.07-0.10). After 1 hr. shaking, indoleacrylic acid (0.5 ml of a 1 mg/ml stock in 100% ethanol) is added (to induce the *trpE* promoter which drives *Turf13*) and shaking is continued for 2 hr at 30 C (OD550=~0.1-0.2). Cells are spread with a glass rod on the surface of L plus ampicillin agar medium (1 ml cells/15 cm petri dish). To assay for toxin in fungal cultures, cylinders of agar medium (4 mm dia.) bearing mycelium are placed mycelium side down on the spread surface and incubated overnight at 30 C. Clear zones in the bacterial lawn surrounding cylinders indicate a positive toxin reaction. More uniform distribution of bacterial cells can be obtained by embedding them in a top agar overlay (induced cells are diluted 1:1 with L plus ampicillin containing 2% agar). Either method of applying cells can be used to test for toxin in solution by adding 5 ul of solution (e.g. culture filtrate or purified toxin preparation) to a filter paper disk lying flat on the agar surface.

Quantitative assays can be performed in liquid medium. Toxin-containing solution is added to cells induced as described above. In a typical assay 100 ul toxin is added to 1 ml cells in a glass test tube (13 x 100 mm). The OD550 is measured, cells are incubated for various periods of time at 30 C, and an OD550 is taken at each time point. Data from a representative assay are shown in Fig. 2. Note that PM-toxin is as readily quantified as T- toxin, that as little as 1 ng toxin/ml can be measured, and that this assay is good for detecting toxin in unprocessed culture filtrates from a Tox+ strain. There is no inhibition of cell growth by culture filtrate from a Tox strain, instead growth is stimulated at high filtrate concentrations.

CHARACTERIZATION OF A MUTATION IN A STRAIN OF PENICILLIUM CAMEMBERTII AFFECTING THE PRODUCTION OF CYCLOPIAZONIC ACID.

Penicillium camembertii is a filamentous fungus used for the production of mold- fermented white cheese. It is a domesticated form of *Penicillium commune*, especially adapted to the food environment (Pitt et al. 1986. Food Microbiol. 3:363-371). Despite its use as food starter culture, *P. camembertii* is able to produce cyclopiazonic acid (CA), a secondary metabolite toxic to animals and humans (Holzapfel 1968. Tetrahedron 24:2101- 2119); Le Bars 1979. Appl. Environ. Microbiol. 38:1052-1055). The synthesis of CA starts from the amino acid tryptophan with the condensation of acetyl-CoA and isoprenoids, especially dimethyl allylpyrophosphate (Holzapfel 1980. In: P.S. Steyn, The biosynthesis of mycotoxins, Academic Press). Acetyl-CoA is also the direct precursor of the isoprenoids.

To define the genes responsible for the production of cyclopiazonic acid, *P. camembertii* Ps 912 was treated with nitrous acid and screened for CA negative mutants. One mitotically stable strain was isolated with only 1-2% of CA production compared to the wild type (Geisen et al. 1990. Appl. Environ. Microbiol. 56:3587-3590).

To localize the mutation either in the pathway for biosynthesis of isoprenoids or in the biosynthetic branch between tryptophan and CA, radioactive labelling either with 3H-tryptophan or with 14C-acetate was carried out. For this purpose *P. camembertii* was grown on minimal medium (per liter: 5 g glucose, 3.75 g KH2PO4, 0.5 g MgSO4, 0.1 g NaCl, 0.1 g CaCl2, 0.75 g KOH, 1.2 g KNO3, 15 g agar) for 5 days at 25 C and transferred to minimal medium containing either L-[5-3H]-tryptophan (at a concentration of 30 uCi/ml with a specific activity of 31.5 Ci/mmol) or [14C]-acetic acid, sodium salt (at a concentration of 10 mCi/ml with a specific activity of 59 mCi/mmol), and grown at 25 C for 5 days. After that time the same amount of cell material either from the mutant strain or from the wild-type strain was recovered from a colony, transferred to a microcentrifuge tube and treated with 1 ml chloroform for 5 min in a microcentrifuge shaker. The mycelium was discarded and the chloroform evaporated in a vacuum centrifuge. The residue was redissolved in 10 ul chloroform and applied to a thin layer chromatography plate (Kiesel-gel 60, Merck, Darmstadt) and chromatographed in two dimensions (solvent systems: first dimension, chloroform/methanol:9/1; second dimension chloroform/isobutylmethylketone:4/1). The Rf values were determined by dividing the migration length of the respective spot by the migration length of the solvent. After chromatographic separation the plates were autoradiographed.

The mutant and the wild-type strain were first labelled with 3H-tryptophan. If there is no regulatory feedback mechanism to prevent an accumulation of the respective precursor there should be an accumulation of a new metabolite just "in front" of the mutation if the mutation is located within the biosynthetic branch between tryptophan and CA. The results of the experiment are shown in Figure 1. Beside the clear signal of CA in the case of the wild type, there was no other chloroform extractable 3H-tryptophan-labelled material under the conditions used. In the case of the mutant strain, there was no signal at all, indicating the mutant had not integrated radioactively labelled tryptophan into extractable precursors of CA. These results indicate the mutation is probably not located within this biosynthetic pathway, assuming there is no regulatory feedback mechanism. The fact that the mutant strain was also able to grow on minimal medium without tryptophan indicates the mutation has not affected the pathway for biosynthesis of this amino acid.

To test the possibility that the mutation has influenced the pathway for isoprenoid biosynthesis, the mutant and the wild-type strain were grown on minimal medium containing 14C-acetate. Acetate is the precursor of the isoprenoids, so labelling with 14C-acetate should give different patterns of extractable metabolites if the mutation is located within this pathway. Figure 2 shows the result of this experiment. There is a different signal pattern between the mutant and the wild type. In the case of the mutant there are clearly differences in the separation of the spots on the baseline of the first dimension. These spots are not able to migrate by using the second solvent system. A clear resolution of one spot in the second dimension at a higher Rf value (0.868) could be observed in the case of the mutant. In the case of the wild type, resolution of two spots occurred.

The solvent system used in the second dimension shows poor separation capacity for these substances. The fact that the Rf values are different in the mutant compared with the wild type (0.723 and 0.568) indicates different labelled substances. The results show that the mutation affects different chloroform extractable metabolites labelled with 14C-acetate rather than affecting a single substance. Similar complex changes of the labelling pattern of the mutant compared to that of the wild type could be observed when the strains were incubated for a longer period of time (14 days or 21 days) on medium containing 14C-acetate (data not shown). In these cases the overall patterns from both the mutant and the wild type were different from the patterns, indicating a change of the accumulated labelled metabolites during the time course of incubation. These results show that the mutation leads to complex changes in the pattern of the 14C-acetate-labelled metabolites and suggests it is located in the pathway for isoprenoid biosynthesis.

Alternative possibilities that condensation of acetyl-CoA residues to tryptophan is abolished or that the mutation affected the pathway from tryptophan to CA, and that an accumulation of a precursor of that pathway is prevented by a feedback mechanism are not likely because in any case the 14C-acetate labelling pattern should not be changed.

NON-OAK RIDGE RFLP PATTERNS IN THE *NEUROSPORA CRASSA MULTICENT-2* STRAINS

The *N. crassa* strain *multicent-2* (FGSC 4488) is widely used for RFLP mapping (Metzenberg et al. 1984 Neurospora Newsl. 31:35-39). The genetic background of *multicent- 2* is considered as being largely "Oak Ridge". However, by studying RFLP patterns, we have recently found that a region in the *multicent-2* linkage group VII differs considerably from that of the standard wild type "Oak Ridge" 74-OR23-1A (FGSC 987) (Haedo et al. 1992 Genetics, in press). We extend here this observation to a 40 kb region on the left arm of linkage group I. Table I shows that the analyzed region on chromosome I is largely "Oak Ridge" both in *multicent-2* and, as expected, the Mauriceville-1c genome.

Although this finding does not modify any conclusion based on the use of *multicent-2* in RFLP mapping experiments, it indicates that a *multicent-2* RFLP or, in general, the length of a *multicent-2* restriction fragment is not necessarily "Oak Ridge", and would not be expected in either Southern blots or DNA clones originating from the current standard wild type "Oak Ridge" strains or libraries, respectively.

METHODS: *N. crassa* DNA was purified, restricted and subjected to Southern blot analysis as described (Haedo et al.

1992 Genetics, in press). A pMOcosX cosmid clone containing a 40 kb insert closely linked to the *arg-3* locus (M. Mautino, S. Haedo and A.L. Rosa, unpublished), on the left arm of linkage group I, was 32P-dATP-labelled and used as a probe. The *N. crassa* wild-type pMOcosX library was constructed by Drs. M. Sachs and M. Orbach and is distributed by FGSC.

NEW ROUND SPORE MUTATIONS IN *NEUROSPORA CRASSA* ACCOMPANYING CHANGES IN A DUPLICATION CLOSELY LINKED TO THE *R* LOCUS.

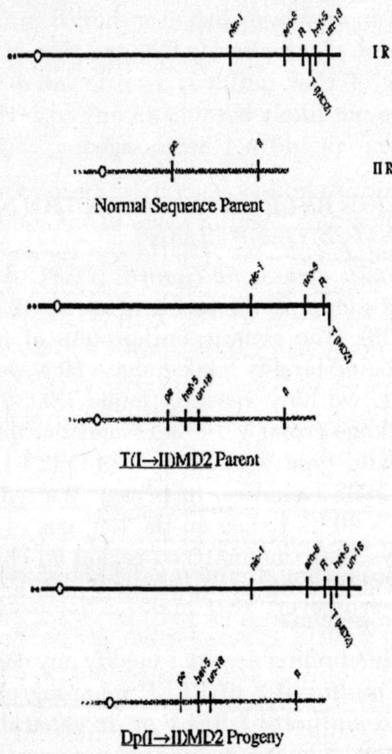

Figure. Partial genetic map of linkage groups I and II of parental normal sequence and *T(I->II)MD2* and *Dp(I->II)MD2* progeny. *het-5* and *un-18* are covered, but *R* is not. Insertion point and orientation in linkage group II have not been determined exactly, although translocation breakpoints have shown linkage to *pe* (Perkins, unpublished).

Various genotypes can result in round ascospores in *N. crassa* (Mitchell 1966 Neurospora Newsl. 10:6; Barry et al. 1972 Neurospora Newsl. 19:17). I have observed that round ascospores can also originate in crosses where one parent is the apparent breakdown of a partial diploid heterozygous at the heterokaryon incompatibility locus *het-5*. The breakdown may remove one of the duplicated chromosome segments carrying a *het-5* allele and possibly extend past the duplication and affect the nearby Round spore locus (*R*).

Crosses of normal chromosome sequence with the insertional translocation *T(I->II)MD2* result in a class of progeny which duplicates the far distal portion of linkage group IR (Figure 1). The duplication covers loci *un-18* and *het-5* but not *R* (Perkins and Jacobson, unpublished). Mutations at *R* are ascus dominant; in asci from *R* X *R*+, all eight ascospores are round. *R* cultures also show abnormal, peach-like vegetative morphology that is recessive in heterokaryons. Noncoverage is inferred from the cross normal sequence *R* X *T(I->II)MD2 R*+; the ratio of *R* to *R*+ progeny (as judged by vegetative morphology) is 2:1 indicating that the duplication progeny, *Dp(I->II)MD2*, are *R* and the locus is outside the duplication.

If normal sequence and *T(I->II)MD2* parents in such crosses have different alleles at *het-5*, *Dp(I->II)MD2* progeny will be heterozygous for *het-5* and will exhibit abnormal restricted colony morphology. After prolonged incubation the colony will escape inhibition and resume growth. (Such escape is common in duplications that are heterozygous for a het gene and it can result from any process that eliminates the heterozygosity, e.g. deletion or mitotic crossovers that make it homozygous.) The morphology of the culture after escape resembles conidial separation mutants in a tap test and is characteristic of *T(I->II)MD2* and of *Dp(I->II)MD2* where *het-5* is homozygous. A proportion of these escaped progeny produced round ascospores in subsequent crosses even though *R* was not present in either parent.

The following set of crosses illustrates this in detail (Table 1). *het-5*PA was introgressed by backcrossing from strain Panama CZ30.6 (FGSC 1311) into Oak Ridge wild type (FGSC 2489 and 4200) which carries *het-5*OR. After the third backcross a *het-5*PA progeny was crossed to *T(I->)MD2 het-5*OR. Twenty-two of 69 progeny showed restricted colony morphology indicating heterozygous duplication of *het-5*. The inhibited *Dp(I->II)MD2* were not stable; all escaped and grew into *T(I->)MD2*-like morphology.

All were then crossed to normal sequence. Only five were barren (the usual result of duplication X normal sequence), while the other 17 were fertile to some extent as judged by production of viable ascospores. Twelve of the fertile escaped *Dp(I->II)MD2* produced only normal shaped ascospores, three produced mixtures of normal and round, and two produced exclusively round ascospores. Each ascus contained either all normal or all round ascospores and each perithecium contained only one of the two types of asci. The three crosses producing both shapes contained a mixed population of normal and round-ascospore producing perithecia.

Progeny were analyzed from the two exclusively round ascospore crosses. No inhibited, presumably heterokaryon incompatible, progeny were recovered. This, combined with the *T(I->)MD2*-like morphology, suggests that the original *Dp(I->II)MD2* lost the *het-5*PA allele when it escaped and either remained a duplication or reverted to translocation sequence. In any case, they did not produce duplications heterozygous for *het-5* in subsequent crosses to normal sequence *het-5*OR.

Three vegetative morphologies were seen among the second generation progeny: wild type (14), *T(I->)MD2*-like (18), and a peach-like morphology which has been associated with R (Barry et al. 1972 Neurospora Newsl. 19:17). Morphology was correlated with the shape of the ascospores produced when these progeny were crossed to normal sequence. If they were fertile, cultures with wild type morphology and *T(I->)MD2*-like cultures both produced only normal ascospores. However, the majority of the *T(I->)MD2*-like cultures were barren and this would suggest that they were newly generated duplications. The peach-like cultures produced only round ascospores. The round ascospore trait was heritable through another cross to normal sequence. Yet another morphological class was apparent among the progeny of this cross. This class exhibited dense mycelium appressed to the agar and produced exclusively round ascospores in crosses to normal sequence.

These new round ascospore strains resemble the known *R* mutant by being ascus dominant, female sterile in heterozygous crosses and completely sterile in homozygous crosses. Crosses between these strains and *R* are also sterile.

Morphology of the MD2-like culture resembles conidial separation mutants in a tap test and is characteristic of *T(I->II)MD2* and of *Dp(I->II)MD2* where *het-5* is homozygous. The morphology of the appressed mycelium class in cross 3 consisted of dense mycelium appressed to the agar.

Ascospore shape produced in crosses of progeny X normal sequence. Wild type is spindle shape; round is spherical similar to *R* mutant; (—) is barren in crosses with no ascospores produced.

*het-5*PA had been backcrossed three generations to normal sequence Oak Ridge wild type (FGSC 2489 and 4200).

& Progeny from the previous generation which produced exclusively round ascospores were used as male parents and individually crossed to normal sequence (FGSC 2489 and 4200). The number of progeny were pooled from all crosses.

The escaped *Dp(I->II)MD2*, or derived progeny, that produced round ascospores could not be classified as either normal or parental *T(I->)MD2* sequence. In each cross, white inviable ascospores were produced in varying proportions. This was true in each of the crosses described above and when the round ascospore strains were crossed to *T(I->)MD2.* Distribution of white ascospores in unordered asci shot from perithecia was inconclusive in determining the nature of chromosome sequence. Direct examination of rosettes also did not reveal patterns characteristic of heterozygous *T(I->)MD2* or known simple rearrangements. Moreover, a portion of progeny in each generation were barren in subsequent crosses. This suggests possible novel chromosome aberrations which may be producing duplication progeny.

A proportion of escaped *Dp(I->II)MD2* from a seventh backcross *het-5* X *T(I->)MD2* also produced round ascospores, confirming the correlation with the escape of heterozygous *het-5* duplication rather than other interactions of Panama and Oak Ridge backgrounds. *Dp(I->II)MD2* homozygous for *het-5*OR was stably barren, thus supporting the conclusion.

Somatic instability of duplications is common when they are placed at selective disadvantage, for example when heterozygous for het genes (Perkins and Barry 1977 Adv. Genet. 19:133-285). The unusual aspect of this study is that a gene outside the duplicated area is apparently affected during breakdown of the duplication. A possible explanation is that the distal portion of the normal linkage group IR is lost during escape resulting in quasi-translocation sequence. This is also unusual since most studied duplication breakdowns yield normal sequence (Perkins and Barry 1977 Adv. Genet. 19:133-285). The lost portion of IR may occasionally extend proximally beyond the *T(I->)MD2* breakpoint and thus affect the *R* locus. Some *Dp(I->II)MD2* which have escaped appear to be mixtures of round and normal ascospore genotypes, suggesting that either loss of distal IR occurs repeatedly in a colony or the chromosome sequence after deletion is itself unstable.

The round ascospore trait became stable after the first cross. It is unclear, however, if the chromosome sequence stabilized after meiosis. The varying proportion of white ascospores, differing colony morphologies, and barren progeny cannot be simply explained and requires further investigation.

These observations were incidental to a study carried out at Stanford University to localize *het-5* right of the *T(I->)MD2* breakpoint. Support from PHS Grant AI-01462 and helpful suggestions from members of David D. Perkins laboratory are gratefully acknowledged.

Role of Amino Acids in Halotolerant Aspergillus Repens

Attempts have been made earlier to understand the mechanisms of osmoregulation and the ability ot adapt to fluctuations in the external osmolarity (Ben-Amotz and Avron 1983. Ann. Rev. Microbiol. 37:95-119, Csonka 1989. Microbiol. Rev. 53:121-147). Various strategies have been adopted by the cells in order to survive and proliferate in the presence of reduced water activity, including accumulation of carbohydrates. amino acids and quaternary ammonium compounds (Measures 1975. Nature 257:398-400, Brown et al. 1972. J. Gen. Microbiol. 72:589-591, Dannibier et al. 1988. Arch. Microbiol. 150:348-357). In the present study, attempts were made to investigate the role of amino acids that accumulate intracellularly in halotolerant Aspergillus repens under salt stress condition.

Aspergillus repens was grown in 50 ml synthetic medium in 250 ml Erlenmeyer flasks. The composition and culture condition was essentially the same as described earlier (Parekh and Chhatpar 1989. Curr. Microbiol. 19:297-301). For amino acid analysis, cell-free extracts were treated with trichloroacetic acid (TCA). After precipitation and centrifugation, TCA was removed from the supernatant by chilled supernatant. The supernatant was then subjected to amino acid analysis by automatic amino acid analyser (LKB) with known standards. Proteases were assayed as described earlier (Chhatpar et al. 1984. Experientia 40:1382-1384) from the cell-free extracts. Glucose-6-phosphate dehydrogenase (G6PDH) and FDP aldolase activity was assayed as described by Chhatpar et al. (op. cit.). Glutamate dehydrogenase (GDH) activity was assayed using the procedure described by Meers and Tempest (1970. J. Gen. Microbiol. 64:187-190).

Significant differences were observed in free amino acid pools at 144 h when the mold was grown in medium supplemented with 2 M NaCl (stress condition) as compared to control growth conditions (Table 1). At 72 hours, however, no or few amino acids were observed. Amino acids like porline, serine and glutamate were found to be higher in mold grown under stress as compared to control conditions. Non-polar amino acids like alanine, valine and leucine were found to be significantly higher than other free amino acids detected. When the osmolytes like glutamate, proline, glycerol and glycine-betaine were added to the growth medium at a concentration of 10 mM, there was an enhanced production of free amino acids under saline conditions.

Analysis of intracellular proteases (acidic, neutral and alkaline, pH optima 5.5, 7.0 and 8.6, respectively) revealed an increase in their activities at all stages of growth under stress as compared to controls. The intracellualr portease activity was significantly higher from 48 to 72 h and thereafter gradually decreased. Other enzymes like FDP aldolase and GDH were also more active uncer stress as compared to controls at 72 and 144 h of growth.

An increase in the amino acid pool as well as an increase in specific amino acids will act in several ways to overcome salt stress problems. They are listed as follows: a) Glycine and alanine have high water solubility and will rescue the cell by overcoming reduced water activity; b) Proline associates itself with its hydrophobic part with the hydrophobic side chain of protein, thereby converting them into hydrophilic groups by exposing the carboxylic and imino groups towards the water molecules providing a proper water structure under reduced water activity (aw); c) Acidic amino acids like glutamate and apsartate have a net negative charge. It is possible that internally accumulated acidic amino acids will sequester sodium and will diminish the excessive positive Na+ charge; d) Osmotic accumulation of K+ is regulated by proline; e) Glycine-betaine and glutamic acid increase with the increasing external osmotic strength and overcome growth inhibition caused by osmotic stress by protecting enzymes from inactivation at high ionic strength.

Decreased levels of free amino acids at 72 h may be accounted for by their rapid utilization for the de novo synthesis of proteins rather than their production through degradation of proteins, since new proteins may be essential for growth under 72 h. However, comparatively more hydrolysis of proteins may be taking place than synthesis resulting in more amino acids accumulating at 144 h of growth under stress as compared to control. Barlow et al. (1976. Crop Sci. 16:59-62) have observed 20 t0 250% increase in free amino acids in Zea under stress. Similar results were obtained in bean leaves (Huber et al. 1977. Z. Pflanzen. Physiol. 31:234-247). The accumulation of significantly higher levels of free amino acids within the cell would draw water back into the cell and thus lower the otherwise deleterious concentrations of intracellular solutes.

In our studies there was not only a net increase in total amino acids under stress but levels of non-polar amino acids like alanine, leucine and valine were found to increase at a higher rate compared to other amino acids. Ben-Naim (1980. Plenum Publishing Corp., New York) has investigated the phenomenom of hydrophobic effect. Non-polar molecules in aqueous solution tend to aggragate, squeezing out water. Although different assemblies of water molecules have different energies, each component of the system has to contribute to the equilibrium i.e. final state, in which water must have the same chemical potential throughout. Water is in a state of low free energy because of high concentration of solutes (here, as NaCl) and thus increase its chemical potential. Water adjacent to the non-polar molecules move apart and decrease their chemical potential. It is possible that non-polar amino acids (i.e. leucine, valine and alanine) in aqueous solution function by decreasing the chemical potential, to control the chemical potential of the cell because of sodium chloride which would otherwise have been deleterious to the cell.

Protein turnover is essential for the adptation of cells to new environmental conditions, and intracellular proteases play a vital role in protein turnover processes, along with protein synthesis. A high level of protease at 48 to 72 h indicates a high protein breakdown and hence the turnover at these hours, but surprisingly, the free amino pool at this stage is very low, suggesting the possibility of more conversion to, than degradation of protein at this stage. The reverse may be true at 144 h where synthetic potential may be comparatively much less than degradation leading to accumulation of amino acids.

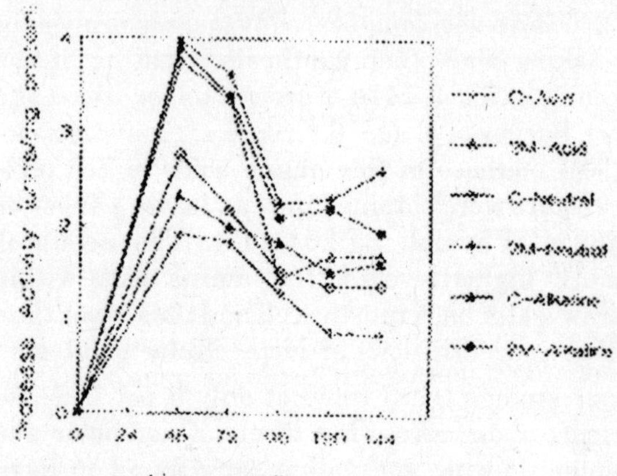

Figure 1. Intracellular protease activity from *Aspergillus* [illegible] grown under saline conditions. 3M = 3 M NaCl supplemented media; C = control medium (no NaCl)

Besides the production of proteases, there was a significant increase in activities of enzymes like FDP aldolase, cytosolic GDH and G6PDH at 72 and 144 h in cells grown under stress as compared to controls. Increase in the activity of FDP aldolase and G6PDH under saline conditions may suggest an increase in the rate of glycolysis and hexose monophosphate shunt (HMP shunt) which will provide intermediates for amino acid biosynthesis. The glycolytic intermediate pyruvate may provide alanine, valine and leucine while 3 P-glycerate would give rise to serine, cysteine and glycine. Phosphoenol pyruvate and erythrose 4-P may produce phenylalanine, tyrosine and tryptophan, while histidine can be produced from ribose. Saline conditions increase GDH activity, which will produce glutamate, resulting again in the production of proline and arginine. Aspratate is produced from oxaloacetate. Aspartate in turn can give rise to isoleucine, lysine and threonine. Thus, saline conditions can enhance the cellular metabolic pathways like glycolysis, TCA cycle and HMP shunt, etc. which may lead to the production of intermediates of amino acid biosynthetic pathways and thus contribute to an increase in the amino acid pool under saline conditions.

In our studies with *Aspergillus repens*, then, total amino acid accumulation, enhanced production of non-polar amino acids and increased levels of proteases and enzymes of carbohydrate metabolism may play crucial roles enabling adaptation to saline conditions.

A *QA-2+*-PGEM VECTOR FOR NEUROSPORA TRANSFORMATIONS

The *qa-2+* gene has been widely used as a selectable marker in Neurospora transformations (Akins and Lambowitz 1985. Mol. Cell. Biol. 5:2272-2278). A plasmid (pMSN1) has been constructed to facilitate the cloning of selected genes into a *qa-2+* vector. The system takes advantage of the blue/white screening possible when cloned fragments are inserted into the *E. coli lacZ* (beta-galactosidase) gene, and also facilitates subsequent sequencing of the cloned genes.

The *qa-2+* gene of *N. crassa* was inserted into a non-coding region (the *Nde*I site) of the parental pGEM-3Zf(+) (Promega) to create the pMSN1 vector. After pGEM-3Zf(+) was digested with *Nde*I, the recessed 3' ends were filled in with the Klenow fragment of DNA polymerase I and the blunt-ended molecules were phosphatase-treated (Cobianchi and Wilson 1987. Methods Enzymol. 152:94-110). An approximately 2.5 kb *Bam*HI fragment (containing the 2.1 kb *Bam*HI-*Hin*dIII fragment of the *qa-2+* gene (Geever et al. 1989. J. Mol. Biol. 207:15-34) plus a 345 bp *Hin*dIII-*Bam*HI fragment from pBR322) was made blunt with Klenow fragment, gel-purified and ligated to the prepared pGEM-3Zf(+) vector. The 5.8 kb pMSN1 plasmid was transferred into NM522 by selecting ampicillin resistant transformants (Miller 1987. Methods Enzymol. 152:145-170).

The pMSN1 vector contains a multiple cloning site within the coding sequences of the *lacZ* gene. When an *E. coli* strain harboring the *lacZ* M15 gene on an F' (such as NM522; Gough and Murray 1983. J. Mol. Biol. 166:1-19) is transformed with pMSN1, those transformants form blue colonies on indicator plates containing IPTG and X-Gal. Strains harboring recombinant vectors with inserts at the multiple cloning site of pMSN1 (which disrupt the *lacZ* gene) form white colonies on indicator plates.

The pMSN1 vector contains SP6 and T7 RNA polymerase promoters that flank the multiple cloning site, allowing the in vitro synthesis of RNA from either strand of the cloned inserts. Also, pMSN1 contains the origin of replication of the filamentous bacteriophage f1, which facilitates the production of single-stranded DNA for sequencing.

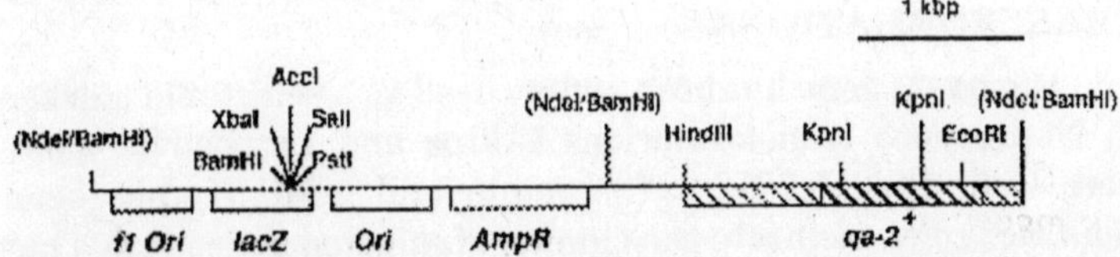

Figure. The pMSN1 vector (approximately 5.8 kb). The circular construct has been linearized at one of the destroyed *NdeI/Bam*HI sites (indicated in parentheses). The open boxes represent regions within the pGEM-3Zf(+) sequence, while the dark-slashed and light-slashed boxes indicate coding and non-coding areas, respectively, with the Neurospora *qa-2+* insert. The positions of the beta-galactosidase gene (*lacZ*), the ampicillin resistance gene (*AmpR*), the origin of replication for growth in *E. coli* (Ori), and the filamentous bacteriophage f1 origin of replication (f1 Ori) are shown. The five unique enzymes (*Bam*HI, *Xba*I, *Acc*I, *Sal*I and *Pst*I) cutting within the multiple cloning in the *lacZ* gene are shown. The multiple cloning site also contains sites for *Eco*RI, *Sac*I, *Kpn*I, *Sma*I, *Ava*I, *Hinc*II, *Sph*I and *Hin*dIII, which cut within the Neurospora sequence (not shown). Selected sites within the *qa-2+* insert are indicated. The plasmid is drawn approximately to scale.

The recombinant vectors are introduced into Neurospora by transformation of *qa- 2;aro-9* spheroplasts and selection for prototrophy (growth in the absence of aromatic amino acid supplement; Akins and Lambowitz 1985. Mol. Cell. Biol. 5:2272-2278).

We examined the progeny obtained with recombinant pMSN1 constructs (Nelson and Metzenberg 1992. Genetics, in press) and showed that nearly half contained single inserts of the transforming DNA sequences. The transformants can be analyzed for recombinant DNA phenotypes and/or used in RIP-mediated gene disruption experiments (Selker et al. 1989. Fungal Genetics Newsl. 36:76-77).

EFFECT OF LIGHT ON CONIDIATION OF *NIT* AND *LIS* MUTANTS OF *NEUROSPORA CRASSA* GROWN ON DIFFERENT NITROGEN SOURCES

In addition to a recent paper (Ninnemann 1991. J. Photochem. Photobiol. 9:189-199), we are reporting further data on *Neurospora crassa* mutants supporting our hypothesis that the flavohemoenzyme nitrate reductase (NR) is involved in photoreception for light- promoted conidiation (Klemm and Ninnemann 1979. Photochem. Photobiol. 29:629-632). We looked for light-stimulated conidiation and NR activities in the NR-regulatory mutants *nit-2* (FGSC 2698), *nit-4* (2992) and *nit-5* (985) from the Fungal Genetics Stock Center, University of Kansas, as well as the four NR+ mutants *bd lis-1* (2891), *bd lis-2* (2892), *bd lis-3* (2983), and *al-2 bd,* received from Dr. S. Brody, University of California, San Diego. The three "light insensitive" *lis* mutants were isolated and described by Paietta and Sargent in 1983 (Genetics 104:11-21): the expression of their circadian rhythm was relatively insensitive to continuous light.

The mutants were grown either on NR-inductive Vogel's medium (with 10 g/l glucose) modified so that 50 mM NaNO3 was the only nitrogen source, and solidified with 1.2% agar, or on repressive medium (same as above with nitrate replaced by 25 mM NH4Cl). In addition, we chose two repressive-derepressive media with either 25 mM NH4NO3 or with arginine (5 g/l) plus 3 mM $NaNO_3$ (plus glucose at 3 g/l for partial starvation) added to the salts of Vogel's medium (Ninnemann 1991. J. Photochem. Photobiol. 9:189-199) so that some NR activity could be induced. The albino-band strains *al-2 bd* in which we had found light-promoted conidiation long ago (Klemm and Ninnemann 1978. Photochem. Photobiol. 28:227-230) was included in all experiments as a control. Maintenance of the strains, the method of growing them in large (20 cm) Petri dishes and irradiating them with white light (5 W/m2, 0.5-24 h) as soon as the mycelia had grown one third of the diameter of the Petri dishes have been described before (Ninnemann 1991. J. Photochem. Photobiol. 9:189-199).

The effect of light on conidiation was recorded photographically (Reimer 1992. Master's Thesis, University of Tübingon). Since conidia cannot quantitatively bo washed off the mycelia for a quantitative determination, the culture plates and the photographs thereof were evaluated qualitatively. This evaluation is summarized in Table 1. Indicated is the lack or appearance of conidia stimulated by light in addition to conidia eventually formed in darkness.

The three regulatory nit mutants (NR-) were blind for light-inducing photoconidiation on the three media tested (nit mutants cannot grow on nitrate medium), although carotenoid synthesis was induced in the irradiated mycelia. The *al-2 bd* control, however, photoconidiated on inductive nitrate medium as well as on medium with arginine and low concentration of nitrate. However, this strains photoconidiated not at all or scarcely on the repressive ammonium medium (no NR activities present) and less profusely on the repressive-derepressive ammonium nitrate medium than on inductive nitrate medium.

Photoconidiation occurred in *bd lis-2* and *lis-3* on inductive nitrate medium; *bd lis-1* grew very poorly on nitrate medium so that no conidia were formed in darkness or in light. No, or scarcely any photoconidiation was observed on repressive ammonium. Conidia were clearly, but not profusely light-induced in the three *lis* mutants grown on the two repressive-derepressive media. Thus conidiation could be photostimulated in the three *bd lis* mutants under conditions where at least some nitrate reductase was induced: the *lis* mutations did not affect photoconidiation. At the same time, the growth of *bd lis-3* and especially of *bd lis-2* was inhibited by increasing time of irradiation (>2 h). Also, carotenoid synthesis was light-induced in these mutants.

MITOTIC INSTABILITY IN BENOMYL-RESISTANT TRANSFORMANTS OF A FLUFFY STRAIN OF *NEUROSPORA CRASSA*

The isolation of the beta-tubulin gene from a benomyl-resistant *Neurospora crassa* strain (Orbach et al. 1986 Mol. Cell. Biol. 6:2452-2461) has provided a dominant selectable marker usable in transformation experiments with *N. crassa*.

In order to determine whether plasmid pBT6 (bearing the benomyl-resistant beta-tubulin gene on the pUC12 vector) stably integrated into genomic DNA, we used it to transform the morphological mutant fluffy of *N. crassa* (FGSC 45). This strain can be conveniently used in transformation experiments because it generates only uninucleate microconidia, allowing the isolation of homokaryons and avoiding the complications in isolation via ascospores (Rossier et al. 1985 Curr. Genet. 10:313-320). The fate of pBT6 DNA was analyzed in transformants grown under selective or non-selective conditions.

To produce microconidia, the fluffy strain was grown for 7 days at 25 C in Petri plates on Westergaard and Mitchell's synthetic medium supplemented with 1% acetate as sole carbon source (acetate medium). Transformation experiments were performed according to Orbach et al. (1986 Mol. Cell. Biol. 6:2452-2461) with modifications. Microconidia were inoculated in Neurospora minimal medium (Difco no. 0817-01) at a concentration of 1 x 10(7) cells per ml. Cultures were incubated 24 hours at 33 C at 150 rpm. Germinating microconidia (about 5 x 10(8)) were washed 3 times with distilled water at room temperature and resuspended in 10 ml 1 M sorbitol to which 2 ml Novozym 234 (Novo Industri A/S, batch 1199, 5 mg/ml in 1 M sorbitol) were added. Digestion was performed for 1 hour at 30 C at 100 rpm. After centrifugation, spheroplasts were washed twice with 10 ml 1 M sorbitol and once with 10 ml 5 mM Tris-HCl buffer pH 8.0, 1 M sorbitol, 50 mM CaCl2 (transformation buffer). Spheroplasts in transformation buffer were diluted with 0.6 volume distilled water just before transformation (Akins and Lambowitz 1985 Mol. Cell. Biol. 5:2272-2278). Five microliters of DMSO and 0.1 ml 50 mM Tris-HCl pH 8.0 containing 40% polyethylene glycol 4000 (PEG) (BDH) and 50 mM CaCl2 were then added to the spheroplasts in 0.4 ml diluted transformation buffer. Ten micrograms of plasmid pBT6 DNA purified twice by CsCl ultracentrifugation in 100 ul TE, pH 8.0, and preincubated 10 min with 25 ul heparin solution (5mg/ml, freshly prepared in transformation buffer), were then added to the transformation mixture. After 30 min incubation on ice, 5 ml PEG-Tris-CaCl2 solution were added and incubation continued for 20 min at room temperature. Appropriate aliquots of the transformation mixture were inoculated with 3% agar, 1 M sorbitol, 2% sorbose, 0.05% fructose and 0.05% glucose) at 45 C. This medium was poured on a 25 ml bottom layer (Vogel's medium N containing 2% sorbose,

0.05% fructose, 0.05% glucose and 1.5% agar to which 0.5 ug/ml benomyl was added after autoclaving). Plates were incubated for 3 days at 33 C.

Spheroplasts derived from microconidia germinated for 24 hours could be transformed with pBT6 with a frequency of 1000 to 4000 transformants per ug DNA. Microconidia incubated for 12 h were still ungerminated and were not suited for transformation due to poor formation of spheroplasts. Homokaryotic derivatives from the presumably heterokaryotic primary transformants were obtained by isolating colonies grown from microconidia inoculated on selective (0.25 ug/ml benomyl) Vogel's medium N supplemented with 2% sorbose, 0.05% fructose and 0.05% glucose.

The proportion of transformed nuclei and their stability in benomyl-resistant transformants were determined by genetic analysis of microconidia produced on selective (0.5 ug/ml benomyl) or non-selective acetate medium, assuming that the microconidia contain a random sample of the nuclei present in the mycelium. The percentage of transformed microconidia was determined by counting the colonies obtained from an equal number of microconidia plated onto selective (0.25 ug/ml benomyl) and non-selective Vogel's medium N (both contained 2% agar plus 2% sorbose, 0.05% glucose and 0.05% fructose). A variable degree of mitotic instability was observed during growth in the absence of selective pressure. Under such growth conditions, the microconidia of some transformants kept the transformed character (35% of benomyl resistant microconidia in transformant 1f) while microconidia of others lost it (<0.3% of transformed microconidia in transformant 3i).

To investigate at the molecular level the effects of the suppression of selective pressure on the fate of the transforming DNA, 3 transformants (1f, 9 and 3i) were grown in Vogel's medium N plus (0.25 ug/ml) or minus benomyl. DNA was then extracted and purified by CsCl ultracentrifugation. Undigested plasmid pBT6 DNA and undigested genomic DNAs from the transformants grown in the presence of benomyl were electrophoresed on 0.8% agarose gels and probed with nick-translated pBT6. No bands were detected that would correspond to the relaxed circular or supercoiled configurations of the plasmid (results not shown). This is consistent with the integration of pBT6 into genomic DNA and with the absence of free, autonomously replicating plasmids.

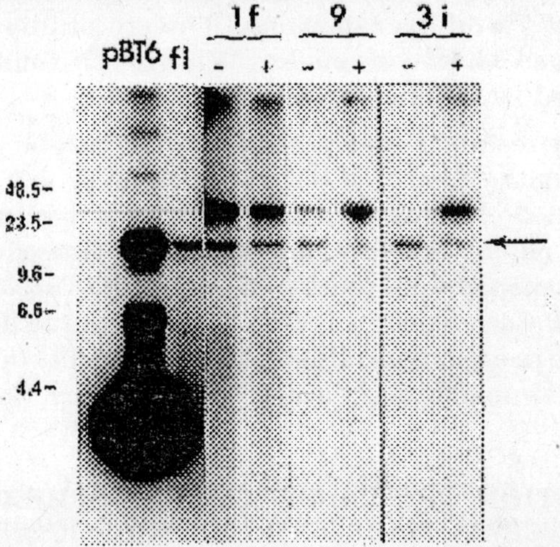

Figure. Genomic Southern blot analysis of DNAs isolated from untransformed strain fluffy (*fl*) of *N. crassa* and from 3 benomyl-resistant transformants (1f, 9 and 3i). Transformants were grown in the absence of selective pressure (-) or in the presence (+) of 0.25 µg/ml benomyl. 1 ug of DNA from each strain and 1 ng of plasmid pBT6 were electrophoresed after digestion with *Cla*I on 0.8% agarose gels and probed with 32P-labelled pBT6. As *Cla*I does not cut pBT6, it migrated mainly in the supercoiled configuration. The arrow points to the 13 kb *Cla*I DNA fragment bearing the resident beta-tubulin gene. Size markers (lambda DNA digested with *Hin*dIII) are shown in kb; *fl*: untransformed fluffy strain.

*Cla*I does not cut pBT6. In untransformed fluffy, pBT6 hybridized to a single *Cla*I DNA fragment of about 13 kb bearing the resident beta-tubulin gene. In all genomic DNAs from the transformants, pBT6 hybridized to this band and to another *Cla*I fragment of higher molecular weight indicating that integration occurred by non-homologous recombination. However, the *Cla*I fragments bearing pBT6 sequences were shorter than the corresponding undigested DNAs, ruling out the presence of non-integrated circular oligomers of pBT6 (date not shown). In the absence of selective pressure, pBT6 sequences were maintained (transformant 1f) or lost (transformant 3i). In transformant 9, the intensity on the autoradiogram of the *Cla*I fragment bearing the pBT6 sequences is weaker after growth in the absence of selective pressure. This suggests that some nuclei lost the pBT6 sequences, in agreement with the finding that only 7% of transformant 9 microconidia kept the transformed character under such growth conditions, as determined by genetic analysis.

These results strongly suggest that mitotic instability observed in the absence of selective pressure is due to excision of integrated plasmid sequences. These observations contrast with the mitotic stability observed in *N. crassa* with other transformation systems (Case 1986 Genetics 113:569-587; Avalos et al. 1989 Curr. Genet. 16:369-372) and with the finding that pBT6 gave rise to mitotically stable transformants in *Podospora anserina* (Fernandez-Larrea and Stahl 1989 Curr. Genet. 16:57-60).

GENERATION OF TRANSFORMABLE SPHEROPLASTS FROM MYCELIA, MACROCONIDIA, MICROCONIDIA AND GERMINATING ASCOSPORES OF *NEUROSPORA CRASSA*

For Neurospora to be generally useful in molecular studies it would be desirable to be able to prepare transformable spheroplasts from mycelia and any of the three types of spores produced by this organism. Transformable spheroplasts are currently prepared from germinating macroconidia by digestion with Novozym 234 in the presence of 1.0 M sorbitol (Vollmer and Yanofsky 1986. PNAS 83:4869-4873). This method is efficient, but requires a 3-5 hr germination step. Elimination of the germination step would be a technical advance. In addition, the standard method is usable only with strains that form large numbers of macroconidia. Thus, interesting mutants that are incapable of forming macro- conidia cannot be used as recipients in cloning experiments.

A procedure for generating spheroplasts from mycelia of *N. crassa* has been reported (Buxton and Radford 1984. MGG 196:339-344). While large numbers of spheroplasts are released by this procedure, the frequency of transformation is low, and we have experienced difficulty obtaining repeatable results. Since we want to clone genes implicated in the macroconidiation process, we devised a procedure that improves the efficiency of transformation of mycelial spheroplasts. As an alternative approach, we developed an transformation protocol for microconidia. Since aconidial mutations can be introduced into a microcycle microconidiating background such as mcm (Maheshwari 1991. Exp. Mycol. 15:346-350), transformation of microconidia represents a viable option for the cloning of conidiation genes. A procedure for generating competent spheroplasts from germinating ascospores also was developed and provides an additional strategy for cloning conidiation genes.

In this paper, we describe protocols which generate transformable spheroplasts from mycelia, macroconidia, microconidia and germinating ascospores. The main alteration from the standard spheroplast formation protocol is the use of MgSO4 (1.0 M) instead of sorbitol (1.0 M) as the osmotic stabilizer. In addition, a higher concentration of Novozym 234 (Novo Laboratories, Wilton CT, lot no. PPM 1530) and a modified culture procedure are utilized for generation of spheroplasts from mycelia. We also report on a modification of the standard spheroplast regeneration procedure which results in more rapid appearance of transformants and a 2-4 fold increase in transformation frequency of spheroplasts from germlings and mycelia.

Mycelia

1. We have used an inoculum build-up procedure to generate large amounts of actively growing mycelia. One hundred milliliters of Vogel's medium plus 1.5% sucrose (Davis and de Serres 1970. Meth. Enzymol. 17A:79-143) is inoculated with either conidia or mycelia from an agar culture. The culture is macerated in a 360 ml Waring blender cup (two 10 s bursts with a 10 s pause), and incubated overnight on a rotary shaker (200 rpm) at 34 C. This primary culture is macerated and 30 ml are used to inoculate a secondary culture (final volume 130 ml). After overnight incubation, the secondary culture is macerated and 30 ml are used to inoculate a tertiary culture. After continued incubation (see below) mycelia from the resulting tertiary culture is centrifuged (2,000 x g, 10 min), resuspended in 30 ml of 1.0 M MgSO4 and recentrifuged.

2. A crude estimate of mycelial wet weight is obtained from the difference in weights of the washed culture in the centrifuge tube and the empty tube. The culture is resuspended in approximately 2 ml of Novozym 234 (10 mg/ml in 1.0 M MgSO4) per gram wet weight of mycelia. Digestion is performed at room temperature with gentle inversion on a Belly Dancer shaker (Stoval Life Science Inc., Greensboro NC) for 1 h. The resulting digest is placed on ice, and all subsequent treatments performed at 4 C.
3. The mycelial digest is filtered by gravity through 6 layers of cheesecloth placed in the bottom of a 12 ml syringe. An equal volume (approximately 15 ml) of ice-cold sorbitol (1.0 M) is passed through the syringe and cheesecloth and mixed with the spheroplast suspension by gentle inversion. Spheroplasts are sedimented at 55 x g for 10 min at 4 C. The resulting pellet is resuspended in 15 ml of ice-cold 1.0 M sorbitol and recentrifuged as above. The pellet is next resuspended in 15 ml of ice-cold STC (1.0 M sorbitol; 50 mM Tris, pH 8.0; 50 mM CaCl2) and recentrifuged. The spheroplasts are finally resuspended at a concentration of 1-2 x 10(7) spheroplasts/ml in spheroplast storage buffer containing STC, PTC (40% PEG 4,000, 50 mM Tris, pH 8.0; 50 mM CaCl2) and dimethylsulfoxide in a ratio of 8:2:0.1. We routinely freeze the resulting spheroplast suspension slowly by placing in a -80 C freezer.
4. Transformations are performed according to Vollmer and Yanofsky (PNAS 1986 83:4869- 4873). We typically mix 10 ul of heparin (5 mg/ml in STC), 1 ug transforming DNA and 100 ul (2 x 10(6)) mycelial spheroplasts, and incubate on ice for 30 min. (Reactions containing larger amounts of mycelial spheroplasts lead to decreased transformation efficiency, even in the presence of increased heparin concentrations). One milliliter of PTC is added to the reaction mixture and incubation is continued for an additional 20 min at room temperature. Ten milliliters of regeneration agar maintained at 50 C are added to the transformation reaction, and the mixture is poured onto plating medium containing either

hygromycin (250 U/ml) or benomyl (1 ug/ml). Transformants

Using the above procedure, we have generated approximately 5 x 10(7) spheroplasts from either 3, 4 or 6 h tertiary cultures of the wild-type strain (74-OR23-1A), and 5 x 10(8) spheroplasts from a 9 h culture. Osmotically sensitive spheroplasts have been obtained from overnight cultures of wild-type, but these proved to be less competent. With mycelial spheroplasts we typically observe regeneration frequencies of about 10%. Approximately 50 stable, benomyl resistant transformants have been obtained from 10(6) spheroplasts using 0.5 ug of either pBT3 or pMO63 (Selitrennikoff and Sachs 1991, FGN 38:90-91). Efficiency of transformation with the hygromycin resistance plasmid pMP6, which contains the hygromycin phosphotransferase gene driven by a modified *cpc-1* regulatory region (M. Plamann, pers. comm.) was approximately 10-fold lower. We have obtained similar transformation frequencies using mycelial spheroplasts generated from the aconidial mutant *acon-3* (Matsuyama et al. 1974 Dev. Biol. 41:278-287).

Macroconidia

1. Macroconidia from a 2-3 week agar culture grown at 22 C are suspended in sterile water, filtered through cheesecloth, and centrifuged for 10 min at 800 x g. The conidia are then resuspended in 30 ml of 1.0 M MgSO4, sedimented as above, and resuspended in 10 ml of MgSO4 per 1 x 10(9) conidia.
2. Two milliliters of Novozym 234 (5 mg/ml in 1.0 M MgSO4) are added per 1 x 10(9) conidia, and the conidial suspension is incubated on a rotary shaker (100 rpm) at 30 C until osmotically sensitive cells are generated (typically 30-60 min). The reaction is placed on ice, and all subsequent procedures are performed at 4 C.
3. An equal volume of ice-cold 1.0 M sorbitol is added to the spheroplast suspension, and the mixture is centrifuged at 500 x g for 10 min at 4 C. The resulting pellet is resuspended in 15 ml of 1.0 M sorbitol and recentrifuged as above. The pellet is next resuspended in 15 ml of STC and recentrifuged. Resulting spheroplasts are finally resuspended at a concentration

of 1 x 10(8)/ml in spheroplast storage buffer. (Some clumping of spheroplasts is typically observed at this step). Aliquots of the competent spheroplasts are dispensed into microcentrifuge tubes and slowly frozen by placing in a -80 C freezer.

4. Transformation reactions are performed as described for mycelial spheroplasts, except that 1 x 10(7) conidial spheroplasts are used per transformation reaction.

Using the procedure described, we typically obtain between 100 and 1,000 hygromycin-resistant transformants using 1 x 10(7) spheroplasts from a variety of strains and 1 ug of pMP6. Approximately 25% of the transformants are stable when transferred to fresh hygromycin medium. Transformation efficiencies using pDH25 and pCSN43 (Staben et al. 1989 FGN 36:79-81), which contain the hygromycin phosphotransferase gene driven by the *trpC* promoter of *Aspergillus nidulans*, were dramatically reduced compared to pMP6. Transformation to benomyl resistance with pBT3 (Orbach et al. 1986 Mol. Cell Biol. 6:2452-2461) was unsuccessful. We have generated transformable spheroplasts from either fresh conidia as described above, or conidia stored in water at 4 C for 1 or 2 days.

Microconidia

Microconidia from *mcm* are generated as described by Maheshwari (1991) Exp. Mycol. 15:346-350) by inoculating 100 ml Vogel's medium + 1.5% glucose in a 500 ml Erlenmeyer flask with 5 x 10(8) macroconidia and shaking at 240 rpm, 22 C for 16-24 hr. After filtration through cheesecloth (which generates a filtrate consisting of >99% microconidia) the microconidia are pelleted by centrifugation (800 x g, 10 min) and resuspended in 50 ml 1 mM EDTA pH 8.0, 25 mM 2-mercaptoethanol. The microconidia are shaken (100 rpm) for 20 min at 30 C and then pelleted by centrifugation (800 x g, 10 min). (It is uncertain whether this pretreatment is necessary for spheroplast formation). From this point on, the procedure described above for spheroplasting and transforming macroconidia is followed (steps 2-4). Spheroplast viability was ~12% and 300-500 transformants were obtained when 1 ug of pMP6 and 10 ul of 5 mg/ml heparin were mixed with 1 x 10(7) spheroplasts.

Germinating ascospores

1. Ascospore production is initiated by inoculating the female parent onto the center of 150 mm petri dishes containing synthetic crossing agar (Davis and de Serres 1970 Meth. Enzymol. 17A:79-143) and incubating at 25 C for 7-10 days with a 12 hr light/dark cycle. Conidia (1 x 10(6)) are added to each plate for fertilization. The plates are then incubated at 25 C for an additional 3 weeks. (Plates are inverted in the dark for the final two weeks). Ascospores are collected off the lids of the petri dishes by scraping in sterile H2O with a rubber policeman and pelleted by centrifugation (800 x g, 5 min). A typical yield is 1 x 10(8) ascospores per plate.
2. Ascospores are resuspended in 10 ml Vogel's medium, transferred to a 250 ml Erlenmeyer flask and activated by heat shocking at 60 C for 45 min. Following this treatment, ascospores are added to 140 ml of Vogel's medium and germinated by shaking at 200 rpm, 30 C for 4-7 h. (We have observed that germination, which occurs from both ends of the ascospore, is not as synchronous as conidial germination). Ascospore germlings are collected by centrifugation (800 x g, 5 min) when 30-75% of the spores have germinated, and germ tubes are 1-5 ascospore lengths.
3. Germlings are washed once in 20 ml ice-cold 1.0 M sorbitol and then resuspended in 10 ml ice-cold 1.0 M sorbitol. Spheroplasts are prepared by adding 2 ml 5 mg/ml Novozym 234 (in 1.0 M sorbitol) and then shaking gently (100 rpm) for 30-60 min at 30 C. (The dewalled germ tubes often detach from the ascospore shell during this step). The spheroplasts are then washed and prepared for transformation as described above for macroconidia starting at step 3.

Using pBT3, 10-900 transformants per microgram DNA per 5 x 10(6) spheroplasts have been obtained. The transformation efficiency appears to vary widely with different parental genotypes. Transformations with plasmids containing hygromycin-resistance genes resulted in very few, if any, transformants.

Alternate Spheroplast Regeneration Protocol

We have developed a modified regeneration procedure which uses sucrose as the osmotic stabilizer and tergitol (Springer 1991 FGN 38:92) to promote restricted, colonial morphology. Bottom agar contains Vogel's salts (1x), sucrose (15 g/L), tergitol (0.005%) and the selection drug. Regeneration top agar (used at 10 ml/plate) contains Vogel's salts (1x) and sucrose at 0.6 M. We have observed a 2-4 fold enhancement in spheroplast regeneration and transformation efficiency using spheroplasts from germlings and mycelia. When identical aliquots from a single transformation reaction containing wild-type germling spheroplasts and pBT3 were plated on the two different regeneration media, 253 +/- 7 (S.E.) benomyl resistant transformants were obtained on the sucrose/tergitol medium, while 61 +/- 10 transformants were obtained using the standard regeneration medium. Dilutions of the same experiment plated on non-selective media resulted in 19.0 +/- 3.2 regenerates on sucrose/tergitol medium and 9.7 +/- 1.7 regenerates on the standard regeneration medium. In addition, transformants were evident after 24 hr at 30 C using the sucrose/tergitol medium, but did not appear until 48 h incubation using the standard regeneration medium. A potential problem with the new method is that the colonial growth morphology is lost when the colonies emerge from the growth medium. As a result, the number of transformants per plate must be reduced, and transformants must be transferred after 2 days at 30 C. The benefit of increased efficiency must be balanced with the less restricted growth morphology. We have not determined the efficiency of the modified regeneration procedure using spheroplasts from conidia or ascospores.

VERSATILE FUNGAL TRANSFORMATION VECTORS CARRYING THE SELECTABLE BAR GENE OF *STREPTOMYCES HYGROSCOPICUS*

Several selectable genes have been reported for construction of filamentous fungal transformation vectors. Among the most widely used is the *hygB* (also known as *hph*) gene of *E. coli*, which is generally useful because the corresponding selective agent (hygromycin B) is toxic to wild type strains of many fungi and because scoring of transformants is usually unambiguous.

We, and others (Avalos et al. 1989 Curr. Genet. 16:369-372), have found that the same merits are evident using bialaphos (or phosphinothricin) as a selective agent and the *bar* gene (DeBlock et al. 1987 EMBO J. 6:2513-2518), which encodes phosphinothricin acetyltransferase, as a selectable marker. We report here the construction of three vectors which carry bar as the selectable gene and have easily exchangeable parts as well as convenient cloning sites.

The first plasmid (pBP1) was constructed as follows. A 575 bp *Bam*HI fragment carrying the bar coding region was inserted into the *Bam*HI site of pUC18 (Yanish-Perron et al. 1985 Gene 33:103). A 631 bp *Sal*I-*Dde*I fragment carrying the *Cochliobolus heterostrophus* Promoter 1 element (Turgeon et al. 1987 Mol. Cell. Biol. 7:3297-3305) was end-filled, attached to *Xba*I linkers, and inserted into the pUC18 *Xba*I site immediately 5' of the *Bam*HI site, thus creating a Promoter 1::*bar* transcriptional fusion (Fig. 1). The second plasmid (pBP1T) was made by inserting a 470 bp *Acc*I fragment (blunt-ended and attached to *Eco*RI linkers) from the 3' untranslated region of the *C. heterostrophus TRP1* gene (Turgeon et al. 1986 Gene 42:79-88) into the *Eco*RI site of pBP1, thus providing a fungal terminator (Fig. 1). The junction regions were sequenced as shown below:

```
-----P1-----/     BamHI               BamHI        KpnI        EcoRI  /-TRP1-
TGCTCAgctctagagGATCCATGAGC---GgatccccgggtaccgagctcgaattccAGAC
     DdeI     XbaI  /-----------bar-----------/        SmaI      SacI
```

Restriction enzyme sites are underlined or overlined. Linker and vector sequences are in lower case letters. The 3' end of the Promoter 1 (P1) fragment is shown fused to the *Xba*I linker. Both ends of the *Bam*HI fragment carrying the *bar* gene are shown (internal sequences are omitted); the start codon is in bold type. The 5' end of the *TRP*1 terminator fragment is shown fused to the *Eco*RI linker.

An important feature of these plasmids is that their relevant parts (promoter, coding region, terminator) can be readily removed or exchanged with other sequences, simply by digestion with the appropriate enzyme (*Xba*I, *Bam*HI or *Eco*RI) and religation with or without a substitute fragment attached to the proper linkers. Note that some of the unused polylinker sites of pUC18 are no longer unique because they are also found in one or more of the inserts. Remaining cloning sites are *Eco*RI, *Sma*I, *Sac*I and *Hin*dIII for pBP1 and *Hin*dIII only for pBP1T.

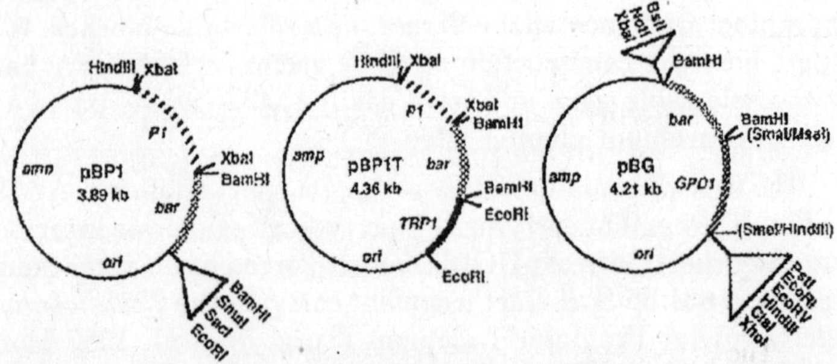

Figure. Construction of plasmids pBP1, pBP1T and pBG. Steps in construction are described in the text. Restriction enzyme sites shown are either unique (for cloning) or points at which the sequences indicated in the text were inserted. *amp* = *E. coli* ampicillin resistance gene; *ori* = *E. coli* origin of replication; *P1* = *C. heterostrophus* Promoter 1; *bar* = *E. coli* bialaphos resistance gene coding region; *TRP1* = *C. heterostrophus* tryptophan biosynthetic gene terminator; *GPD1* = *C. heterostrophus* glyceraldehyde-3-phosphate dehydrogenase gene promoter.

The third plasmid (pBG) was made by inserting the 575 bp *Bam*HI *bar* fragment into the *Bam*HI site in the polylinker of the Bluescript vector pIIKS+. A 675 bp *Hin*dIII-*Mse*I fragment bearing the promoter of the *C. heterostrophus GPD1* gene (VanWert and Yoder 1992 Curr. Genet. in press), was end-filled and inserted into the *Sma*I site of pIIKS+, just 5' of the *bar* gene, thus creating a *GPD1* promoter::*bar* transcriptional fusion (Fig. 1). A combination of sequencing and restriction enzyme analysis confirmed the junction regions as shown below.

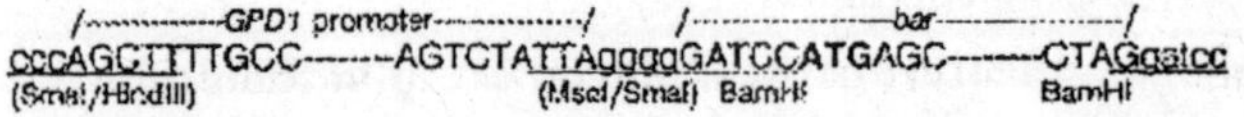

Conventions are as described above for pBP1T. Restriction enzyme sites in parentheses are nonfunctional.

All three plasmids were used to transform *C. heterostrophus*, using standard procedures (Turgeon et al. 1987 Mol. Cell. Biol. 7:3297-3305). The protoplast regeneration medium for selection of transformants was modified to contain only osmoticum and Cochliobolus minimal salts (Leach et al. 1982 J. Gen. Microbiol. 128:1719-1729), solidified with 1% agarose and containing either bialaphos or phosphinothricin at a final concentration of 50-100 ug/ml. Complex media were avoided since phosphinothricin (a synthetic compound also known as glufosinate-ammonium, an analog of L-glutamic acid) specifically inhibits glutamine synthetase. Bialaphos, a naturally-occurring tripeptide consisting of phosphinothricin and two residues of L-alanine, is toxic to cells after it is converted to phosphinothricin by endogenous cellular peptidases which remove its L-alanine residues.

The transformation frequency with each of the plasmids was 1-10 fast-growing and 50-500 slow-growing colonies/ug plasmid DNA, comparable to the frequencies obtained using similar plasmids but with the *hygB* gene substituted for *bar*. Integration of either pBP1 or pBP1T into chromosomal DNA occurred at both Promoter 1 and at ectopic sites. Single and multiple plasmid copies were observed at either type of site. When transformants were crossed to wild type, the bar gene segregated as a single mendelian element, indicating that integration occurred at a single site in each case. pBP1 and pBP1T were also used to transform *Colletotrichum graminicola*, using procedures similar to those described for *C. heterostrophus*.

PROCEDURE FOR MASS-PRODUCTION OF *NEUROSPORA TETRASPERMA* ASCOSPORES

Heterokaryotic A + a cultures of *N. tetrasperma* produce ascospores more quickly and abundantly than A x a crosses in heterothallic species such as N. crassa. In the course of studying ascospore structure and physiology in the 1950's and 1960's, methods were developed that enabled us to obtain gram-quantities of air-dry spores. Details of the procedure never were published and it has been suggested that I do so now for the benefit of those who may want to obtain ascospores in large quantities.

Our ascospore "factory" was set up with 8-ounce medicine bottles, bought by the case, as the container for growing the spores. Their shape, from above, is as in the diagram below.

We used such bottles, which happened to be brown, because they have a large surface area, and are cheap and easily stacked. It surprised us how few of these non-Pyrex bottles broke in autoclaving during our long use of them.

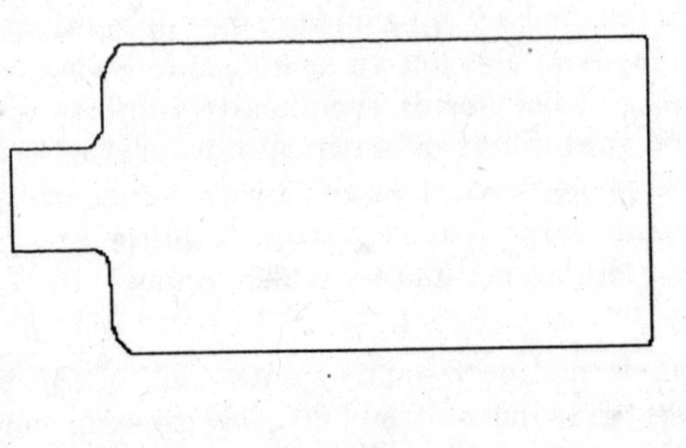

The medium (PYM) we used was as follows: 6.0% potato dextrose agar, 0.5% yeast extract, 0.5% malt extract. Usually, we used Difco materials.

Each bottle contained 50-60 ml of PYM, cotton stoppers were used, and autoclaving was done in the usual way. The bottles were allowed to solidify before inoculation was performed. The medium should not be allowed to dry at room temperature for more than two days before use, and I suspect that in climates with low relative humidity drying should not proceed for more than one day. Therefore, bottles with medium should be stored in the cold if they aren't used quickly, except that this leads to condensation that, in turn, could induce contamination and to more difficulty in seeing what's inside the bottles.

Inoculation was performed with 3-4 ml of heavy suspensions of conidia. It is important to spread the inoculum over the entire surface of the flask because adding enough inoculum is essential to permit spreading by rolling the liquid along the surface. The latter technique avoids the difficulties caused by opening the bottles and using a spreader.

Incubation was accomplished by keeping the cultures in the dark at 25-27 C, with the agar side at the bottom for 1-2 days, after which the agar side should be placed uppermost.

Harvesting was done after 30-32 days, which, under our conditions, yielded the best results. We added 60-100 ml of distilled water to the first bottle and used a flat plastic or rubber scraper to move the ascospores into suspension. This fluid was poured into the next bottle and the process repeated until the suspension got too heavy. This usually happened after about 15 bottles had been rinsed. At that point, the suspension was poured into a powder funnel lined with copper screening (1/32" mesh) which rested in the mouth of a 500 ml or 1 liter cylinder or flask. When the cylinder was full, we replaced it with an empty one and let the spores settle in the original one. After the settling occurred, the supernatant was decanted and the cylinder used again as above. We usually rinsed the bottles a second time using the same technique. It helps not to loosen the agar during this process in order to avoid getting pieces of it, mycelium, and scraps of perithecial walls that settle with the spores and block the pores of the screen.

Cleaning the ascospores was accomplished by repeated (we usually used 5 cycles) settling, decanting and resuspension to remove unwanted materials. Then the spores were shaken in 0.8 EDTA at pH 6.8-7.0 on a reciprocal or rotating shaking machine in a cold room. The EDTA should be changed at two-day intervals for about a week. Upon changing the EDTA solution, carefully decant the suspended matter each time. Several changes of distilled water should be used to replace the EDTA when the shaking is concluded.

This process gave us excellent germination and seemed to get rid of residual contaminants, including conidia. Storage of spores is possible indefinitely at 4 C. The spores can be air-dried and stored on filter paper or in small bottles but germination is more rapid and complete when they are stored in water for at least a week.

NZYME-BASED DNA EXTRACTION FROM ZOOSPORES OF RUMINAL FUNGI

We report here a rapid, efficient and simple method for the extraction of high molecular weight DNA from zoospores of *Neocallimastix frontalis* EB188. This anaerobic fungus, isolated from bovine digesta, effectively degrades plant fiber in vitro (Barichievich et al. 1990. Appl. Env. Micro. 56:43-48). Our interest in ruminal fungi stems from their ability to degrade wood materials (Joblin et al. 1989. FEMS Micro. Lett. 56:119-

122) and their potential use in biomass saccharification. Zoospore DNA synthesis is of particular interest to our laboratory. It is these motile zoospores which colonize and degrade plant materials (Mountfort 1987. FEMS Micro. Rev. 46:501-508). To detail fully this metabolic event, it will be necessary to extract nucleic acids from zoospores. Such procedures have not been reported in the literature. Using acetone drying and enzymatic removal of cell walls, we have isolated high molecular weight DNA from very small amounts of culture. The procedure takes less than one hour and DNA yields are high. The DNA is readily cut with restriction endonucleases and religated efficiently but is otherwise stable. Electrophoretic analysis of the DNA confirmed the presence of repetitive sequences. This procedure will aid the study of DNA replication, DNA repair and DNA RFLP analysis of various strains using small (<1 ml) cultures.

All chemicals, unless otherwise stated, were purchased from U.S. Biochemicals (Cleveland, OH, USA) and were the best grade available. Zymolase (#100,000) was purchased from Seikagaku Kogyo Co. Ltd. (Tokyo, Japan) and was a gift from A. Schroeder. The Lysing Enzyme (#L-2265) preparation was purchased from Sigma Chemical Co. (St. Louis, MO, USA). Restriction endonucleases and T4 DNA ligase were purchased from New England Biolabs (Beverly, MA, USA).

The ruminal fungus, *Neocallimastix frontalis* EB188 was isolated from bovine digesta and grown as described (Barichievich et al. 1990. Appl. Env. Micro. 56:43-48). The strain used (#76100) was obtained from the American Type Culture Collection (Rockville, MD, USA). Zoospores were separated from cultured rhizoidal mass using filtration through 8 layers of cheese cloth and collected using centrifugation at 10,000 x g for 10 min. Zoospores were free from rhizoidal or sporangial debris and counts determined using a Fuch/Rosenthal hemacytometer and microscope. Cultures of *N. frontalis* EB188 released few zoospores (less than 200/ml were present) until 60-63 h after starting cultures. Zoospore numbers increased rapidly afterwards and by 88 h achieved a maximum of 158,000 +/- 11,000/ml. Less than 1200 +/- 400/ml zoospores were found in cultures after 110-120 h. Large (and adequate) numbers of culture zoospores were therefore attainable at 60-110 h.

Zoospores (typically 1 to 2 million) rinsed in water in Eppendorf tubes were resuspended into 100 ul of 0.1 M Na Citrate, pH 7.5, containing 0.5 M sorbitol and 0.06 M Na2-EDTA. When appropriate, zoospores were rinsed in acetone (100%) before enzyme treatments. Acetone treatment (Heath et al. 1986. Appl. Env. Micro. 51:1138-1140) included resuspending the rinsed zoospores into acetone (100%) (0-1 C) and holding on ice for 1-2 min. Zoospores were then collected by centrifugation at 5000 x g (at 4 C) for 2 min. Pelleted material was dried under N2 gas and finally resuspended in enzyme buffer. beta-mercaptoethanol (from a 10% v/v stock) was added to 0.05% (v/v), and Zymolase (or Lysing Enzyme) was added to 80 æg/ml. The sample was mixed and placed at 30 C for 5-10 min. Enzyme protein was determined using the dye binding assay (Bradford 1976.

Anal. Biochem. 72:248-252) with BSA as standard. After incubation, Na-sarkosyl (from a 10% w/v stock) was added to 0.5% (S/v) and proteinase K (from a 10 mg/ml stock solution which had been autodigested several hours at 37 C) was added to 25 ug/ml. The sample was then placed at 62 C for several min. Phenyl methyl sulfonal fluoride (PMSF) (from a 10 mM stock in ethanol) was then added to 0.1 mM and the sample held at 20 C for 1-2 min. Diethyl pyrocarbonate (DEPC) (from a 10% v/v stock in ethanol) was then added to 0.1% v/v and incubation at 20 C was continued for 1-2 min. Tris-HCl, pH 8.0 (from a 1 M stock) was then added to 10 mM. An equal volume (100 ml) of chloroform/isoamyl (25:1 ratio) was then added and the tube gently mixed for 1-2 min to achieve a fine emulsion. Centrifugation at 10,000 x g (at room temperature) for 2 min was used to break the emulsion. The aqueous layer was carefully collected using a trimmed (200 ul) pipet tip. The sample volume was then increased to 500 ul using water. Sodium acetate, pH 6.0 (from a 3 M stock) was added to 0.25 M and isopropyl alcohol was added to 35% (v/v) final.

The sample was gently mixed and chilled to 0-1 C (water/ ice bath) for 2-3 min or held at room temperature for 10 min. Centrifugation at 10,000 x g (at 4 C or room temperature for 5 min was used to collect the precipitate. The pellet was rinsed with 100 ul of absolute ethanol (4 C) and respun at 10,000 x g (at 4 C) for 1 min. The pellet was finally dried under N2 gas and dissolved into a small volume (50 ul) of 10 mM Tris, pH 7.5, containing 1 mM Na2-EDTA. Where necessary, and before gel analysis, ribonuclease A (RNase A) was added (from a stock which was boiled for 10 min.) to 10 U/ml and samples placed at 37 C for 5 min. Quantification of total DNA in zoospores or extracted DNA was performed using the diphenylamine assay (Richards 1974. Anal. Biochem. 57:369-376). Calf thymus DNA was used as standard. Values reported represent the means of triplicate samples. Restriction digestion and ligation of DNA was as suggested by suppliers, and commercial buffers were used. Agarose gel electrophoresis was performed as described (Maniatis et al. 1982. Cold Spring Harbor Lab. Press, p. 545).

Zoospores each possess 2.2 +/- 0.6 pg of DNA. Extraction of high molecular weight DNA from enzymatically weakened zoospores (collected from cultures less than 110 h old) was possible. The median molecular weight of the extracted DNA was greater than the 49 kb marker. Based on ultraviolet absorbance determinations (A230/A260=2.5-3.1), DNA preparations contained large amounts of carbohydrate. Less protein (A230/A260=1.8-1.9) contaminated the DNA preparations. The yield of DNA extracted from zoospores was 61% +/- 7%.

RNA contaminated all DNA preparations but was effectively removed using RNase A. The extracted DNA was stable for long incubations (several hours at 37 C or several days at 4 C) while in a variety of enzyme buffers. Deoxyribonuclease I (DNase I) completely degraded the extracted DNA and *Hin*dIII readily digested the DNA. A variety of other endonucleases including *Pst*I and *Eco*RI also readily digested the DNA. Using sequential incubations (within the same tube), *Hin*dIII digested zoospore DNA was religated efficiently with T4 DNA ligase. *Hin*dIII digested lambda phage DNA could also be effectively religated in the presence of *Hin*dIII digested zoospore DNA. Fragments representing lambda phage DNA could be regenerated (digested) readily from such samples.

Notes:

A. Without Zymolase (or Lysing Enzyme) treatment of zoospores, high molecular weight DNA was not successfully extracted. The use of Zymolase was preferred since it was less expensive. Much lower amounts (less than 10%) of DNA was extracted from zoospores collected from cultures older than 120 h. Such DNA was also degraded (<20 kb) and was unstable (autodegraded) at 37 C. Moderately high molecular weight DNA (about 30-40 kb) in moderate yields (about 30%) could be recovered from overmature zoospores if acetone pretreatments were used before cell wall removal. This extracted DNA was digestible with *Eco*RI and *Hin*dIII. Significantly less RNA was isolated from overmature zoospores, and RNase A treatment was seldom necessary. Evidently, overmature zoospores (which microscopically appeared shrunken) presented cell walls which would not readily lyse (or only partially) without acetone drying (or disruption) of membranes.

B. The proteinase K treatment was important for DNA extraction. Yields of DNA were decreased if this step was excluded. Subsequent PMSF treatment of samples was necessary to inactivate proteinase K since DNA, from several extractions where PMSF was excluded, could only be incompletely digested with *Hin*dIII or *Eco*RI. The use of DEPC during the DNA extraction was necessary since in several preparations (of the more than 50 performed) the DNA was unstable at 37 C. This suggested that nuclease remained in these samples. SDS could be substituted for the Na-Sarkosyl during the lysis step, but additional isopropyl alcohol precipitations (at room temperature) were necessary before the DNA would digest completely with *Hin*dIII, *Eco*RI or DNase I.

C. We have used this procedure to extract DNA successfully from as few as 25,000 zoospores (160 ul of culture fluid from 88-h cultures). Such samples contained less than 0.02 mg dry weight (based on measurements of much larger samples) and provided DNA sufficient for several lane loadings. We have also successfully extracted DNA from 36 mission zoospores (the total of one 250-ml culture), and the extraction was done in two 1.5 ml Eppendorf centrifuge tubes. The DNA digested readily with *Hin*dIII or *Eco*RI and apparently possessed repetitive DNA sequences. This confirmed an earlier report of ruminal fungus DNA extracted from rhizoidal tissue (Brownlee 1989. Nucl. Acids Res. 17:1327-1335). The DNA extraction procedure described here has been used successfully on polyflagellated zoospores from several different (yet untyped) ruminal fungi. This procedure, with or without acetone, could not be successfully used with rhizoidal tissues from either early (65-80 h) or hold (120-160 h) cultures.

IDENTIFICATION OF THE FRAGRANT STRAIN ATCC 46892 AS *NEUROSPORA SITOPHILA*

Neurospora strains were among the various fungi isolated by Park et al. (J. Ferment. Technol. 60: 1-4, 1982) from naturally fermented cassava used to make the indigenous alcoholic beverage tiquira in Maranhao State, Brazil. One strain, deposited as ATCC 46892 and also available as FGSC 6673, was shown by Yoshizawa et al. (Agr. Biol. Chem. 52: 2129-2130, 1988) to have a pleasant fruity odor attributed to ethyl hexanoate. This strain has been the subject of numerous studies concerning production and application of the substance, including pilot studies of the possible use of 46892 in Japan for producing koji and saki (Yamauchi et al. Agric. Biol. Chem. 53: 821-825, 1989). I obtained the strain and crossed it to standard species testers of *N. crassa* and *N. sitophila*. The crosses to both *A* and *a* mating types of *N. crassa* were infertile, but the cross to an *N. sitophila A* tester, FGSC 5940, gave abundant asci with black viable ascospores, showing that 46892 is conspecific and that it does not carry the *N. sitophila* element. Spore killer-1.

LOCATION OF A MUTATION RESISTANT TO COBALT AND NICKEL IN LG IIIR OF *NEUROSPORA CRASSA*

cor is resistant to 10 mM Co2+ or 10 mM Ni2+. It was induced in ORS-6a (FGSC 4200) by serial transfers on cobalt-containing (32 mM) agar medium as described by Venkateswerlu and Sivarama Sastry (Biochem J. 132:673-680). A cross of *cor* to *alcoy* indicated a location in linkage group III. It was then crossed to *acr-2 trp-1 dow*. 98 spores germinated and gave the following results:

CHROMOGENIC AND FLUOROGENIC SUBSTRATES FOR ASSAYING XYLANASES OF NEUROSPORA

The Oak Ridge strains of wild type *N. crassa* grow well with high molecular wight xylan from oat spelts (Sigma X-0376) as the sole carbon source. Xylan, a substituted ß-1,4 linked polymer of xylose, induces high levels of xylanase, xylosidase and ß-galactosidase activities in both culture medium and mycelium. To assay for xylanase activity, chromogenic and fluorogenic substrates have been prepared by procedures based on those of Biely et al. 1985. Anal. Biochem. 144:142-146; Biely et al. 1985. Anal. Biochem. 144:147-151; Rinderknecht et al. 1967. Experientia 23:805; and De Belder and Granath 1973. Carb. Res. 30:375-378. Batches of xylan that have a tan color and granular texture should be dissolved in boiling water, precipitated with two volumes of absolute ethanol, and dried as described below before use. These techniques may also be used to prepare chromogenic and fluorogenic substrates from dextrans, glucans or other high molecular weight polysaccharides to assay endoglycosidases for which suitable assay reagents are not readily available.

Preparation of Blue Xylan (BX) Disperse 2 g xylan in 60 ml deionized water with a magnetic stirrer at room temperature. Ad 0.5 g Remazol Brilliant Blue (RBB) (Sigma R-8001) and stir until dissolved. Add 20 ml of a 1 mg/ml solution of Na2SO4 dropwise over a period of 2 min. Then add 5 ml deionized water and 15 ml of a 10% (w/v) solution of NaOH to initiate the coupling reaction. At the end of 90 min of stirring at room temper-ature, add two volumes of absolute ethanol (200 ml) to precipitate the conjugated xylan. Let stand without stirring for 15 min at -20°C to complete the precipitation. The dyed xylan is collected on a vacuum filter and washed repeatedly with buffered 67% ethanol (1 volume of 0.05 M sodium acetate buffer, pH 5.4 and two volumes of absolute ethanol) until the filtrate is free from all trace of blue color. Break up clumps of precipitate on the filter if necessary. This washing step is crucial as any trapped dye will give an unacceptably high blank. When all traces of blue color are gone from the filtrate, rinse the BX on the filter with 80% ethanol in deionized water. Continue to dehydrate the xylan by washing with 95% ethanol, absolute ethanol and then acetone. Dry under vacuum overnight, pulverize the BX in a mortar, and store at room temperature. Blue xylan made by this procedure has a degree of substitution between 0.08 and 0.10 moles of dye per mole of xylose.

Preparation of Fluorescein labelled Xylan (FX) This procedure is nearly the same as that for BX except that it has been scaled down due to the higher cost of the labelling reagent, DTAF, 5-(4,6-dichloro triazin-2-yl) amino fluorescein hydrochloride (Sigma D-0531). The 6-DTAF isomer may also be used. Deviations from the BX procedure are noted. Disperse 1 g xylan in 30 ml deionized water. Alternatively, the xylan may be dissolved by heating the water to at least 70°C and cooling to room temperature before proceeding. Add 0.10 g DTAF and 10 ml of Na2SO4 as above. Bring to pH 10 with 10% NaOH and add NaOH as needed to keep reaction mixture near pH 10 over the next 2 h at room temperature. At the end, add sufficient deionized water to bring the final volume to 50 ml. Precipitate the labelled xylan with two volumes of absolute ethanol (100 ml) as before and wash on a filter until all of the yellow color is gone from the filtrate. The final traces of free label may be visualized by alkalizing aliquots of the filtrate with NaOH and viewing the fluorescence. Dehydrate the FX and dry in vacuo as above. Fluorescent xylan made by this procedure has a degree of substitution of about 0.02.

Xylanase Assays. These assays directly measure the depolymerization of the high molecular weight substrate and are thus relatively insensitive to the presence of exoxylanases and xylosidases. The assays are run in a 1 ml volume consisting of 0.1 ml of sample and 0.9 ml of a reaction mix containing 5 mg/ml of BX or 2 mg/ml of FX in 0.05 M sodium phosphate buffer at pH 7.5. After incubation at 37°C, the undigested labelled xylan is removed by precipitation with 2 volumes of cold ethanol, chilling for 15 min at -20o C and centrifugation.

The solubilized dyed oligoxylosides are measured spectrophotometrically at 590 or 660 nm or fluorometrically for fluorescein (max 493 nm, Emax nm). Reagent blanks consisting of sample and assay buffer are essential. Other buffers and condition may be used. These assays are quite sensitive to the ionic strength of the assay buffer since the alcohol solubility of the BX and FX fragments varies with buffer concentration.

The absorbance of RBB is relatively unaffected by pH, but the fluorescence of DTAF should be measure above pH 7. Alcohol quenches the fluorescence somewhat so fluorescein standards should be prepared in 67% EtOH.

Localizing Xylanase in gels. BX may also be used to localize xylanase activities in non-denaturing PAGE slab gels. Separating gels are rinsed with 0.01 M sodium phosphate buffer, pH 7.5 and placed in contact with gels consisting of 2% purified agar (i.e. Ionagar #2) and 10 mg/ml of BX in 0.01 M sodium phosphate buffer, pH 7.5. The BX/agar gels are wrapped with Saran wrap to prevent drying and incubated at 34°-37° for several hours. The zymograms are then developed by soaking the BX/agar gels in buffered 67% ethanol (one volume 0.05 M sodium acetate buffer, pH 5.4 and two volumes of absolute ethanol). Enzyme bands show as clear zones against a blue background.

A PRACTICAL METHOD FOR THE PREPARATION OF TOTAL DNA FROM FILAMENTOUS FUNGI

M.I. Borges, M.O. Azevedo, R. Bonatelli, Jr., M.S.S. Felipe and S. Astolfi-Filho Departamento de Biologia Celular, Universidade de Brasilia, 70910 Brasilia, and Departamento de Genetica, Universidade de Campinas, 13100 Campinas, Brazil

Most methods of DNA preparation from fungi are time-consuming due to the need to first make protoplasts, expensive for chemicals such as cesium chloride, or suitable only for small scale preparations. We have developed a simple method for total DNA preparation, yielding a product of quality suitable for restriction digestion and library construction.

To extract DNA, 0.5 to 2.0 g of fresh or frozen mycelium is ground in liquid nitrogen. The mycelial powder is divided between two Sorvall tubes each containing 15 ml of cold spermidine-SDS buffer (4 mM spermidine, 10 mM EDTA, 0.1 M NaCl, 0.5% SDS, 10 mM ß-mercaptoethanol, 40 mM Tris-HCl pH 8.0), and thoroughly shaken. The mixture is then immediately extracted with 1 vol double distilled phenol. After a second phenol extraction, the aqueous phase is extracted with 1 vol chloroform. To the resulting aqueous phase is added 10% of that volume of 3M sodium acetate pH 5.5. DNA is then precipitated by addition of 2 vols of cold absolute ethanol, and recovered either by spooling it out or by centrifugation at 10,000 x g for 10 minutes. After washing with 70% cold ethanol, the DNA is air-dried at room temperature, redissolved in TE buffer (20 mM Tris-HCl pH 7.5, 0.1 mM EDTA) to a DNA concentration circa 0.5 mg/ml and stored at -20°C. Purity and average fragment size are checked by electrophoresis.

The method is highly reproducible, and results over 70% recovery of good quality DNA (A260/A280 circa 1.85). With optimal grinding and nuclease inhibition, the DNA is of high molecular weight (>23 kb), with minimal smear on electrophoresis. The method has been used for *Neurospora crassa, Aspergillus nidulans, Dactylium dendroides* and *Humicola grisea.* For all four species, the DNA prepared has been successfully restricted and religated. DNA from this method has been used in the construction of genomic libraries from *Humicola grisea* and *Dactylium dendroides.*

PRESENCE OF DOUBLE STRANDED RNA IN NATURAL ISOLATES OF *PYCNOPORUS CINNABARINUS*

Cifuentes, V., C. Martinez and G. Pincheira Laboratorio de Genetica, Dpto. de Ciencias Ecologicas, Facultad de Ciencias, Universidad de Chile, Casilla 653, Santiago, Chile.

While studying the nucleic acids of different strains of *P. cinnabarinus* (Pc58, Pc470.3, Pc470.6), the presence of dsRNA molecules (1.7 kb to 3.6 kb) was detected in two of them.

Pycnoporus cinnabarinus is a basidiomycete capable of degrading lignocellulose and is known as a white rot fungus. *P. cinnabarinus* strains Pc58, Pc470.3 and Pc470.6 were isolated from carpophores collected in the rain forest of southern Chile. In the laboratory these strains were grown in liquid PC medium for 15 days at 30°C without shaking. The mycelia were harvested on Whatman #1 paper and washed with distilled water. Then the mycelia were frozen at -20°C and ground in a mortar. Nucleic acids were prepared by the rapid lithium chloride method (Leach et al. 1987 FGN 34:32-33), except that three phenol:chloroform:isoamyl alcohol (25:24:1) and two chloroform:isoamyl alcohol (24:1) extractions were performed before precipitating with ethanol. Nucleic acids were resuspended in sterile distilled water and later were dialyzed against TE buffer (10 mM Tris-HCl, 1 mM EDTA) for 16 h. The preparations were than analyzed by electrophoresis on a 0.7% agarose gel followed by staining with ethidium bromide.

The electrophoretic analysis showed that strain Pc58 has only chromosomal DNA. Strains Pc470.3 and Pc470.6 have four bands moving faster than chromosomal DNA (Fig. 1). These bands correspond to dsRNA as determined by different tests with nucleases. These dsRNAs, denominated as L-dsRNA, M1-dsRNA, M2-dsRNA and S-dsRNA, have been purified by electrophoresis in low-melting-point agarose followed by CTAB extraction (Langridge et al. 1980 Anal. Biochem. 103:264-271). Their molecular sizes were approximately 3.6 kb, 2.7 kb, 2.3 kb and 1.7 kb respectively.

P. cinnabarinus strains Pc470.3 and Pc470.6 seem to have the same types of dsRNA. Furthermore, the presence of different dsRNA in the same strain has been observed as in other fungi (Myers et al. 1988 Curr. Genet. 13:495-501).

Since dsRNA is an indicator of the presence of viruses in fungi, our observations in *P. cinnabarinus* suggest the presence of a virus-like particle which does not seem to be associated with a visible alteration in phenotype of the fungus.

A Method for the Preparation of Intact Chromosomal DNA of *Pycnoporus Cinnabarinus*

Cifuentes, V., C. Martinez and G. Pincheira Laboratorio de Genetica, Dpto. de Ciencias Ecologicas, Facultad de Ciencias, Universidad de Chile, Casilla 653, Santiago, Chile.

We describe an improved procedure to obtain chromosomal DNA of *Pycnoporus cinnabarinus,* which was used in pulse field gel electrophoresis employing a contour-clamped homogenous electric field (CHEF). Four bands of chromosomal DNA were separated. The procedure may be useful to study the genome of other filamentous fungi.

P. cinnabarinus is a basidiomycete capable of degrading components of lignocellulose in nature. Due to the difficulties of conducting classical genetic experiments in this white rot fungus, its genome is totally unknown. Therefore it is interesting to study its DNA through pulse field gel electrophoresis.

The *P. cinnabarinus* strain Pc58, isolated from carpophores collected in the rain forest of southern Chile, was grown for 8 days at 30oC in solid PC medium (Vogel's medium, with a concentration of nitrogen equal to 1/3 of the standard composition, supplemented with 0.2% malt extract, 10 g/ml thiamine, 0.4% glucose and 2% agar-agar).

Intact chromosomal DNA of *P. cinnabarinus* strain Pc58 was prepared by a method based on embedded agarose (Schwartz et al. 1984 Cell 37:67-75). 10(10) spores were suspended in 40 ml of PC medium. They were pelleted at 5000 rpm in a SW40 rotor (Sorvall) for 10 min at 4°C, washed twice with the same medium and germinated for 4 h at 30°C in a rotary shaker at 120 rpm. The germinated spores were pelleted at 5000 rpm in a SW40 rotor for 10 min at 4°C and washed twice with 50 mM EDTA pH 7.5, then suspended in 500 µl of 50 mM EDTA pH 7.5. After addition of 75 µl of Zymolase 100T (Seikagaku Kogyo Co., Tokyo, 10 mg per ml 10 mM sodium phosphate containing 50% glycerol) and 100 µl of Novozyme 234 (20 mg per ml 50 mM EDTA pH 7.5), cells were incubated at 37°C with 1.5 ml of 1% low-gelling-temperature agarose (in 125 mM EDTA pH 7.5), prewarmed to 50°C and then cooled to 40°C.

The suspension was incubated for 30 min at 37°C, distributed in 100 µl molds (LKB cat. 2015-404), and allowed to solidify for 30 min at 4°C. The agarose inserts were added to 4 ml of LET buffer (0.5 M EDTA, 10 mM Tris pH 7.5, 7.5% 2-mercaptoethanol) plus 200 µl of Zymolase 100T (10 mg/ml) and 200 µl Novozyme 234 (20 mg/ml) and incubated for 18 h at 37°C. The inserts were incubated in NDA (10 mM Tris, 0.5 EDTA, 1% lauroyl sarcosine [pH 9.5]) plus 2 mg/ml of proteinase K for 24 h at 50°C. This solution plus proteinase K was replaced once and incubated overnight at 50°C. Inserts were washed twice with 50 mM EDTA pH 7.5 and stored in 50 mM EDTA pH 7.5 at 4°C.

CHEF gel electrophoresis was performed in a LKB pulsaphor, using a hexagonal electrode (LKB cat. 2015-203) and 0.8% agarose gels in 0.1 M TBE. The voltage used, switching pulse and the running times of electrophoresis are described in Fig. 1. Gels were stained with ethidium bromide (1 µg/ml for 60 min and destained in distilled water for 10 min. Intact chromosomal DNA of Saccharomyces cerevisiae strain AB 1380 was used as a standard.

The results of electrophoresis showed four bands of chromosomal DNA (Fig. 1). Densitometric analyses suggest that one of them migrated as a single band, another migrated as a doublet, and the other two bands migrated as a triplet and a quadruplet respectively. The method described here for preparing DNA from *P. cinnabarinus* and its analyses by using pulse fields gel electrophoresis is an alternative way to karyotyping this fungus. In addition, it should be useful for studying other filamentous fungal species.

SEQUENCING OF THE NON-TRANSCRIBED SPACER REGION OF THE RIBOSOMAL RNA GENES OF *NEUROSPORA CRASSA*

The nontranscribed spacer (NTS) region of the ribosomal RNA (rRNA) genes is the most important region of the rDNA because it contains the nucleotide sequences that trigger and/or terminate transcription. To understand the structural organization of the NTS region of *Neurospora crassa* rDNA we have cloned and sequenced the complete NTS region and compared it with the corresponding sequences from other organisms.

One 3592 nucleotide long *N. crassa* nuclear DNA containing the nontranscribed spacer region and part of 26S rRNA (3' end) and 17S (5' end) was isolated and cloned in the plasmid pBR325 (Chambers, C., S.K. Dutta and R.J. Crouch 1986. Gene 44:159). The clone, namely pCC3400, is resistant to ampicillin and tetracycline and sensitive to chloramphenicol.

The complete NTS region of *N. crassa* rDNA (comprising 3592 nt), determined by the chain termination method of nucleotide sequencing, shows many significant features and a brief account of these features follows:

1. Like other rDNA sequences in the NTS region (e.g. Xenopus, human, rat, mouse etc.) the NTS region of *N. crassa* is GC rich (65% G+C).
2. In the NTS region the rRNA processing site 6 is present (indicating that the clone pCC3400 contains the 3' end of the coding region of the 26S rRNA and flanking

sequences) and the sequence GGTGCGAGAACCCGGA (nucleotide residue 226 to 240) is similar to the consensus sequence reported for the rRNA processing site 6 for yeast, Xenopus, mouse and human (but not for Drosophila where the transcription of the tandem array of ribosomal DNA does not terminate at any fixed point.

3. The NTS region contains long stretches of pyrimidines all over, and this type of structural organization if present in maize, wheat, *Raphanus sativus*, *Vicia faba*, pea, cucumber and rice.
4. The transcription termination site, starting with the conventional "*SalI* box" is present in the NTS region of *N. crassa* as has been reported for mouse and human. However, compared to 8 *SalI* boxes in mouse rDNA in the NTS region, only one *SalI* box is present in *N. crassa* (from nucleotide 1469 to 1477). Sequence comparison of the *SalI* boxes and neighboring sequences from different sources suggest that motif TCGAC is common to all.
5. The sequence motif CTTCCT (from nucleotide residue 512 to 517) shows similarity with the human transcription termination site T-2 of its pre-rRNA.
6. The overall sequence of the NTS region of *N. crassa* is closer to that of the *Saccharomyces cerevisiae* NTS region that to those of human, Xenopus, wheat, rice, cucumber, *Vicia faba*, mouse, rat and Drosophila.
7. In the NTS region of *N. crassa* certain sequences like TCTC, TTTT and TTGC reiterate several times. Similar repetition of sequences has been reported for human rDNA although the functional significance of these sequences is not yet established.

Briefly, the information gained in the present studies should be very useful in studying the regulation of transcription of rRNA genes of *N. crassa* and the conserved sequences in the NTS region might be exploited in the evolutionary studies. Determination of the *N. crassa* transcription initiation and termination sites *in vivo* and *in vitro* using clone pCC3400 characterized in the present studies await the results of the appropriate experimental design and as such are a continuous focus of our laboratory. The entire sequence is being published elsewhere and is available upon request.

VECTORS FOR CONSTRUCTION OF TRANSLATIONAL FUSIONS TO BETA-GALACTOSIDASE

The use of *lacZ* as a reporter gene requires appropriate plasmid constructions to produce fusions of fungal promoter and translations initiation sequences to the *E. coli lacZ* gene. I have constructed three plasmids which may facilitate construction of *lacZ* fusions.

The three plasmids are identical except for a *Bam*HI site adjacent to a unique *Sma*I site in three different reading frames with respect to *lacZ*. Thus any blunt-ended fragment containing a functional promoter and translation start site can be used to produce an in frame translational fusion by choosing the appropriate plasmid vector.

The vector contains *E. coli lacZ* from pMC1871 (Shapira et al. 1983, Gene 25:71-82) followed by the terminator of *Aspergillus nidulans trpC* obtained from plasmid DH25 (Cullen et al. 1987, Gene 57:21-26). It also contains the truncated *his-3* gene from pH303 (see accompanying paper) which allows targeted integration of the *lacZ* fusions at the *his-3* locus in FGSC strain 462. The vector for these plasmids is pSP72 (available from Promega Corporation, Madison, WI).

Figure 1 shows a map of pDE1. The plasmid is approximately 9.5 kb in size. The vector sequence of pSP72 runs from the *Xho*I site to the *Sma*I site in all three plasmids. The *Bam*HI site was used to clone the *Bam*HI to *Xba*I fragment from pDH25 containing the *trpC* terminator. The *Hin*dIII to *Sma*I fragment containing the *his-3* gene segment from pH303 was cloned into the pSP72 polylinker at *Hin*dIII to *Pvu*II. Finally the *Bam*HI fragment containing *lacZ* was cloned into the *Bam*HI site adjacent to the *Sma*I site of the pSP72 polylinker. The *Bam*HI site at the *lacZ/trpC* terminator junction was destroyed by cleavage with *Bam*HI and filling of the 5' overhangs followed by ligation. Plasmids pDE1, pDE2 and pDE3 differ at the *Bam*H1 site adjacent to the *Sma*I site. The plasmid with this *Bam*H1 site left intact was designated pDE3. pDE3 was digested with *Bam*HI and ligated following i) mung bean nuclease digestion to generate pDE1 or ii) Klenow fragment of DNA polymerase and dNTP's to generate pDE2.

The nucleotide sequence of this region was determined on both strands to verify the sequence shown in Fig. 1 for each plasmid. The arrows show the cleavage point of *Sma*I. The codon triplets of *lacZ* following the *Sma*I site are also shown. The unique sites in the plasmids are shown in bold: *Bgl*II, *Sma*I and *Hin*dIII. pDE3 also contains the unique *Bam*H1 site adjacent to *Sma*I.

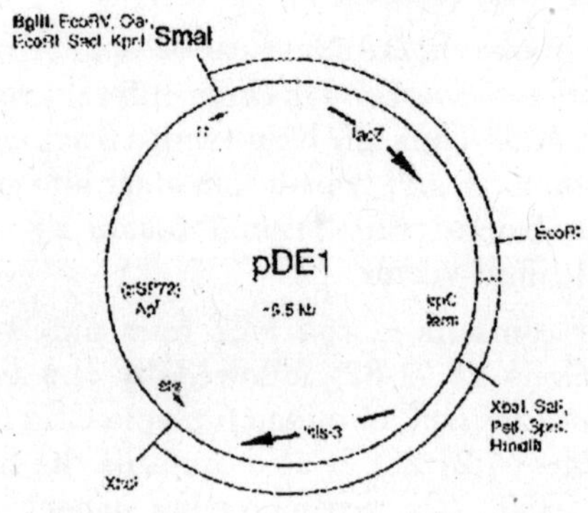

KpnI SmaI
pDE1-- GGTACC CGG GCC GTC GTT TTA--lacZ
pDE2-- GGTAC CCG GGG ATC GAT CCC GTC GTT TTA - lacZ
pDE3-- GGTACCC GGG GAT CCC GTC GTT TTA-lacZ

The *lacZ/trpC* terminator can be removed as a cassette with *Bgl*II (or *Sma*I) and *Hin*dIII for transfer to new plasmids (for example, a plasmid conferring benomyl resistance). Another feature of the plasmid is the T7 and SP6 promoters for synthesizing RNA from each direction for making probes for Southern analysis. The "T7" sequencing primer and *lacZ* "-40" primers can be used for sequencing fragments cloned into the plasmids for verification of constructions.

The *lacZ* gene of pDE3 is in frame with an upstream ATG in the vector sequence and potentially could be used to make transcriptional fusions from promoters cloned into the *Bgl*II site; however, the ATG has a poor sequence context for translation in *N. crassa*. The possible limitation of using these plasmids in the low transformation frequencies obtained by requiring integration at the *his-3* locus in FGSC strain 462. These plasmids have been deposited with the Fungal Genetics Stock Center. This work was performed in the laboratory of Dr. Charles Yanofsky.

A RAPID AND SIMPLE METHOD FOR ISOLATION OF *NEUROSPORA CRASSA* HOMOKARYONS USING MICROCONIDIA

Current approaches for obtaining homokaryons of asexually growing *N. crassa* rely on serial passages of macroconidia, which are usually multinuclear (Davis and de Serres 1970 Methods Enzymol. 17A:79- 143). This method is time consuming and labor intensive. We have devised a simple method for purifying viable uninucleate microconidia from conidiating strains of *N. crassa* grown on Westergaard and Mitchell synthetic crossing medium (SC) supplemented with iodoacetate (IAA). Rossier, Oulevey and Turian (1973 Arch. Mikrobiol. 91:345-353) showed that standing liquid cultures containing 1x SC and 1 mM IAA produced microconidia. We have modified their method to reliably obtain microconidia from 150 mm slants of solid agar medium (0.1 x SC/0.5% sucrose/2% agar/1 mM IAA). We have purified these microconidia free of macroconidia and mycelia using Millipore Durapore Millex 5 µm filters. We are using this technique for the one step purification of homokaryons following DNA-mediated transformation. These methods may be generally useful for studies with heterokaryons.

Experimental

Sterile 16 x 150 mm glass culture tubes containing 6 ml of a sterile, molten solu-tion consisting of 2% agar/0.5% sucrose/ 0.1 x SC and plugged with pre-autoclaved foam plugs, were adjusted to 50-60°C in a water bath. Then 60 µl of 0.1 M sodium IAA (Sigma catalog # I2512, freshly prepared, filter-sterilized in water) was mixed with the contents of each tube (final concentration 1.0 mM). Tubes were slanted at room temperature until solidified, and stored refrigerated for up to one week.

Slants were inoculated using an agar plug of mycelia (obtained by coring a colony with a sterile pasteur pipet) or using a drop of macroconidia suspended in water. Cultures were incubated at 25°C with a 12 hour light/dark cycle for 7-10 days. The production of microconidia was checked by lightly scraping an area of the sparse surface growth (away from tufts of macroconidia) with a sterile, wet inoculating loop. The loopful of culture was transferred to 20 µl of water and examined microscopically.

Microconidia were harvested from cultures by adding 2.5 ml of sterile water to tubes followed by rigorous vortex mixing for 30-60 sec (more microconidia were obtainable by repeating this harvesting step). The conidial suspensions were passed through 5 µm Millex Durapore filter units (Millipore catalog number SLSV025LS) using sterile conditions. Typically 0.1-1% of the microconidia were recovered. The yield as determined by counting with a hemacytometer varied from 10(3)-10(6) microconidia per slant. Filtrates with low numbers of microconidia were concentrated by pelleting the microconidia in a clinical centrifuge (Beckman RT6000, 2000 x g, 5 min). Most of each supernatant was removed by aspiration and the pellets resuspended in the remaining liquid.

Microconidia were germinated at 34°C after spreading on freshly prepared Vogel's/sorbose agar plates. Microconidial viability varied from 1-20%, as determined by comparison of the number of colony forming centers/ml after seven days growth to the number of microconidia/ml as determined by direct counting. We typically plated 2000 microconidia/plate to obtain homokaryotic cultures, and picked colonies after 2-3 days. Vegetative homokaryotic stocks were obtained by transferring individual colonies to slants of Vogel's sucrose medium.

Results and Discussion

We analyzed purified microconidia microscopically using the fluorescent dye Hoechst 33258 to stain nuclei (Springer and Yanofsky 1989 Genes. Dev. 3:559-571). Purified microconidia were almost all uninucleate, although microconidia containing two nuclei and enucleated microconidia were also seen. In contrast, macroconidia from Vogel's minimal slants mostly contained 2-5 nuclei, and macroconidia from the slants of 0.1 x SC/0.5% sucrose/2% agar/1 mM IAA mostly contained 1-3 nuclei.

Genetic analyses of microconidia were consistent with the nuclear counting data. Microconidia were obtained from the forced heterokaryon *cyh-1 ad-3B a*m1 + *al-1 lys-4 a* grown on minimal medium.

These results demonstrate that this method effectively resolved this heterokaryon into its individual ade-requiring and lys-requiring components. The single colony obtained on the minimal plate was subsequently shown to be heterokaryotic in origin, based on analysis of it microconidia.

We have been routinely using microconidia to obtain homokaryons of strains transformed with exogenously applied DNA. Homokaryotic His+ transformants of His- strains were readily obtainable in a single step by selecting for the growth of His+ microconidia on unsupplemented media. Drug resistant transformants could be identified by subsequent screening of cultures grown nonselectively, or by plating microconidia directly on drug-supplemented plates. Microconidia derived from hygromycin B resistant strains transformed with several hph expression constructs readily formed colonies when germinated on plates containing hygromycin B. When plated on benomyl-containing medium (1 μg/ml), microconidia derived from a benomyl-resistant homokaryon germinated poorly (25% control value) and the colonies obtained grew slowly. Seventeen control colonies obtained by germination of the microconidia on drug-free plates were subsequently tested for benomyl resistance: all were benomyl resistant, as expected.

We find that IAA is essential for increased production of microconidia although the mechanism by which it acts is not known (Rossier et al. 1973). We have not examined whether IAA can be autoclaved in the media. When comparing SC concentrations we found 0.1 x SC to give more reproducible results than 0.5 x or 1.0 x SC. We have not tested production of microconidia on slants containing any additional supplements. Growth at 34°C appears to reduce the number of microconidia produced. We have not tried lower growth temperatures, or tested whether the light/dark cycle is necessary for microconidial development.

Nylon 66 (Schleicher and Schuell) with 5 μm pore size can also be used for this procedure but we have more limited experience with it. We don't know precisely the features responsible for the differential filtration of conidia. The poor yield of microconidia suggests that the separation by these membranes may involve properties in addition to pore size. In general, the filtered microconidia germinate over a period of 1-3 days. However, it is prudent to wait for 3 days to be sure that when a colony is picked it is not contaminated with a late germinating microconidium. Microconidia germinate equally well at 34°C or 25°C.

Conclusion

We believe from our preliminary observations that this technique can be of immediate use to Neurospora workers. We have routinely used it successfully to isolate homokaryons from primary transformants and have verified that we have obtained homokaryons by Southern analyses. Using our procedure we can pick a colony from a transformation plate and in 10 days obtain a homokaryon using only one 150 mm slant, one filtration step, one plate, and very little of the investigator's time.

8

Isolation of Fungal DNA for Dot Blot Analysis

A rapid method for the detection of transforming sequences in a fungal strain would be advantageous when trying to determine if unselected sequences are present. Colony hybridization protocols for filamentous fungi have been developed (Stohl and Lambowitz, 1983. Anal. Biochem. 134:82-85; Paietta and Marzluf, 1984.

Neurospora Newsletter 31:40) and a modified method thereof was described (McClung and Dunlap, 1988. Fungal Genetics Newsletter 35:26-27). In this report a simple method for the isolation of fungal DNA from a single transformed colony suitable for dot blot analysis is described. If the original transformant colony is not too small it is sufficient to extract the DNA of one half of that colony. That means that no subculturing is necessary and the results can be available within 24 h starting from the transformant colony.

1. Transfer one half of the transformant colony (or approximately 0.1 g of mycelium from a liquid culture) into a microfuge tube.
2. Add 1 ml of lysis buffer (50 mM EDTA, 0.2% SDS) and 0.1 g of Alumina (Type A5, Sigma).
3. Mix in a microfuge tube mixer (for this protocol an Eppendorf micro tube mixer was used) for 30 min.
4. Centrifuge at 5000 rpm for 5 min.
5. Transfer the supernatant (750 μl) to a new microfuge tube.

6. Extract once with phenol, phenol/chloroform and chloroform each.
7. Add 2 M sodium acetate to a final concentration of 0.2 M.
8. Precipitate the DNA with an equal volume of isopropanol.
9. Dry the pellet and redissolve in an appropriate volume of TE buffer.

With this DNA isolation procedure a contransformation experiment using *Penicillium nalgiovense* ATCC 66742 as host organism was checked by dot blot analysis. The vector p3SR2 which carries the *amdS* gene (Hynes et al. 1983. Mol. Cell. Biol. 3:1430-1439) as a marker was used as a selectable plasmid. As a nonselectable plasmid, pELN5-lac was used. pELN5-lac carries the *E. coli* ß-galactosidase gene under the control of the promoter of the *oliC31* gene (Ward and Turner 1986. Mol. Gen. Genet. 205:331-338) and the terminator of the *trpC* gene (Mullaney et al. 1985. Mol. Gen. Genet. 199:37-45) from *Aspergillus nidulans*.

One microgram of each of the two plasmids were cotransformed and positive transformants were selected for the presence of the *amdS* marker by growth on acetamide minimal medium. The DNA of one half of the resulting transformant colonies (diameter approximately 1 cm) was isolated as described above. Under these conditions 0.1-0.5 µg of DNA was isolated. The whole amount of DNA was used for dot blot analysis. Hybridization was carried out using a digoxygenin labelled DNA fragment which carries the *E. coli* ß-galactosidase gene. Figure 1 shows the result of this dot blot analysis.

The amount of DNA isolated from a single colony gives clear positive signals. In this experiment a contransformation frequency of 70% was reached. After initial screening with the described method, single spores of the cotransformants were subcultured on acetamide minimal containing 0.1 µg/ml X-gal (5-bromo-4-chloro-3- dinolyl-ß-D-galactopyranoside). All strains with positive dot blot signals showed, in contrast to the untransformed strain, a blue color of the mycelium indicating the presence of the *E. coli* ß-galactosidase gene. These results suggest that the cotransformed DNA which was detected by the described method was not in an abortive stage.

The advantage of the described method is the fact, in contrast to the method of McClung and Dunlap (1988. Fungal Genetics Newsletter 35:26-27), which starts from cultures on agar slants as source material, that the original transformed colony can be analyzed for the presence of specific DNA sequences. There is no need for the preparation of protoplasts and the background due to unspecifically bound hybridization probe sequences in very low.

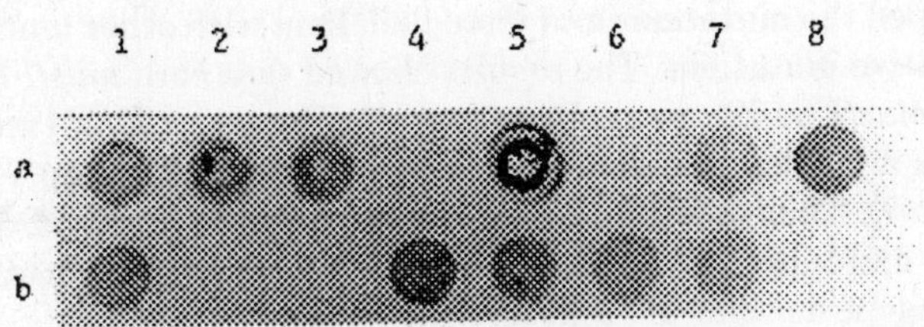

Figure 1. Dot blot analysis of the DNA of *Penicillium nalgiovense* cotransformants. The DNA of 14 cotransformants of *Penicillium nalgiovense* ATCC 66742 transformed with equimolar amounts (1 µg each) of p3SR2 and pELN5-lac was isolated as described in the text. The DNA was resuspended in 100 µl TE, heated at 95°C for 10 min and cooled immediately on ice. The DNA was transferred to nitrocellulose filters (Schleicher & Schull, BA 83) using a Schleicher and Schull Minifold I filtration device. The filter was baked for 2 h at 80°C and prehybridized for 2 h at 68°C. A 3.0 kb DNA fragment carrying a part of the *E. coli lacZ* gene was digoxygenin labelled as described by the manufacturer (Boehringer, Mannheim) and used as a hybridization probe. Hybridization was carried out at 68°C overnight. The washing and developing procedure was performed as described by the manufacturer. The dots 1-7, a+b represent the analyzed samples. As a positive control (8a) 10 ng of pELN5-lac and as a negative control (8b) the DNA of the untransformed strain was used.

DESIGNATION OF NEWLY IDENTIFIED MUTAGEN-SENSITIVE MUTATIONS IN *NEUROSPORA CRASSA* AND THEIR LINKAGE DATA

Two years ago, Dr. Käfer and Dr. Schroeder independently summarized genetic information on mutagen-sensitive mutants in *Neurospora crassa* (Fungal Genetics Newsletter 35:11-13, 1988; DNA Repair 3:77-98, 1988). So far, 30 genetic loci related to mutagen-sensitivity have been indentified. The mutants are more sensitive to UV, X-rays and/or chemical mutagens than the wild-type strain. We have studied DNA-repair mechanisms of Neurospora for several years. During the study, several mutagen-sensitive mutants were isolated and the mutations were mapped. To confirm that they are mutations in new genes, we tested allelism between these mutations and previously reported mutagen-sensitive mutations.

In Fungal Genetics Newsletter 35:11-13, 1988, *mus(FK131), mus(FK132)* and *mus(FK133)* were not yet mapped. We have mapped the mutations and tested allelism with other mutagen-sensitive mutations. The results showed that both *mus(FK131)* and *mus(FK132)* are allelic to *mus-21*. The *mus(FK133)* mutant has a mutation in a new gene, located in linkage group VI (8% to *ylo-1*). Thus *mus(FK133)* has been designated *mus-39*. These new mutants will be characterized for mutagen-sensitivity, mutagenesis, meiosis and fertility.

A QUICK RNA MINI-PREP FOR NEUROSPORA MYCELIAL CULTURES

Most RNA isolation techniques currently in use have been developed for the processing of large quantities of material. These typically involve multiple phenol extractions (Reinert et al. 1981 Mol. Cell Biol. 1:829-836) or guanadinium isothiocyanate/cesium chloride gradients (Chirgwin et al. 1979 Biochem 18:5294-5299) and can be both expensive and time consuming. Often, however, needs arise where quantitatively smaller amounts of RNA are needed from many different samples, for example, during time series analyses or when screening transformants for expression of a transformed gene. Under such circumstances, existing tech-niques are overly time consuming and yield more RNA than is necessary.

The availability of a rapid RNA mini-prep is thus desirable. Such a system has been developed for isola-ting plant RNA (Nagy et al. 1988 Plant Molecular Biology Manual, B4; ed. Gelvin and Schilperoort, Klewer Academic Publishing, pp. 1-29), and we have adapted this procedure for use with Neurospora and, potentially, other filamentous fungi. Below, we describe the use of this procedure with 50 ml mycelial cultures, although we have used in with equal success with 5 ml cultures without scaling down the amounts of any reagents.

The method involves the use of a triphenylmethane dye, aurintricarboxylic (ATA), to protect the RNA. ATA binds irreversibly to RNA and is a potent inhibitor of most nucleic acid binding enzymes (Hallick et al. 1977 Nucl. Acids Res. 4:3055-3064). Thus, RNA made with procedure cannot be used for *in vitro* transcription or translation or reverse transcription but works fine for RNA/DNA or RNA/RNA hybridizations.

To minimize RNase contamination, all glassware is baked at 182°C for at minimum of four hours. Work with gloved hands. The procedure is as follows:

1. Conidia from slants (grown in 16 x 150 mm test tubes containing 8 ml of solid medium) are resuspended in 50 ml of Horowitz complete medium (Horowitz 1947 J. Biol. Chem. 171:255-262) and the cultures grown overnight with shaking at 30°C. A 50 ml culture typically yields enough RNA for 200 gel lanes (see below), and, as noted, smaller culture volumes may be used.
2. Flat mycelial pads are easier to grind than mycelial balls. Therefore, filter cultures using a Buchner funnel onto Whatman #44 filter paper. Wrap flat mycelial pads in aluminum foil and freeze in dry ice. Do not freeze in EtOH/dry ice bath because alcohol might seep through foil. Pads can be stored at -70°C for at least three weeks.
3. Wash a mortar and pestle thoroughly with warm water and Alconox (Fisher Scientific); cool by filling with liquid N2. Remove frozen, flat mycelia from foil and add it to the liquid N2 in mortar. Add ~0.5 g of baked sand and grind mycelial pad to a fine powder. Add more N2 as needed. Mortar and pestle should be washed after every sample.
4. Working quickly before powder can thaw, pour or spoon ground mycelia into 15 ml round bottom Sarstedt tube (Sarstedt tubes #60.540, Princeton, NJ) containing 8 ml of E buffer at room temperature. [E buffer: 50 mM Tris-Cl pH 8.0, 300 mM NaCl, 5 mM EDTA, pH 8.0, 2% SDS; autoclave and add 1 mM ATA and 14 mM ß-mercaptoethanol. ATA=aurintricarboxylic acid, ammonium salt (Sigma #A0885, St. Louis, MO)]

5. Thaw the powder in E buffer in 42°C water bath, occasionally shaking, to get SDS into solution. This should take about 5 minutes.
6. Add 1.1 ml of 3M KCl, invert to mix, keep on ice for 10 min. Solution should form semi-solid, flocculent mass as K-SDS precipitate forms.
7. Spin at 3000g, 4°C in a fixed angle rotor. Make sure caps are screwed on tightly to prevent tubes from collapsing.
8. Pass supernatant through 50 micron Miracloth (Calbiochem #475855, La Jolla, CA) in a funnel into fresh Sarstedt tube.
9. Measure volume of average-sized sample. Add 0.5 vol. 8 M LiCl, mix and stand at 4°C overnight.
10. Spin at 12000g, 4°C, for 15 min in a fixed angle rotor. Thoroughly resuspend pellet in 4 ml sterile gd (glass distilled) H2O with pasteur pipette.
11. Extract twice with phenol/chloroform/isoamyl alcohol (25:24:1), spinning 12000g 10 min at 4°C in a fixed angle rotor. Save aqueous (upper) phase; add gd H2O if volume is less than 2 ml.
12. Add 0.1 vol 3M NaOAc pH 6.0, mix and add 2.5 vol EtOH, mix. Place at -20°C overnight or -70°C for 15 min.
13. Spin 12000g, 10 min, 4°C. Wash pellet with 70% EtOH and drain. Pellet should be a light pink or white.
14. Resuspend in 0.4 ml sterile gd H2O in a microfuge tube. Precipitate with NaOAc and EtOH as in step 12.
15. Spin 10-20 min in microfuge. Wash twice with 70% EtOH and dry pellet. Resuspend in 200 µl of sterile RNase-free gd H2O. Store at -70°C for up to three months. Spectophotometric quantification may be done at this point.
16. Load 1 µl onto a formaldehyde gel. Electrophorese overnight at 20 volts. Blot onto nitrocellulose. Probe with DNA fragment of choice. Expose to film.

Yields are typically on the order of 1-2 mg total RNA from an overnight 50 ml culture arising from an average slant. The number of samples able to be processed using this procedure is limited by the number of spaces in a centrifuge rotor. We have done as many as 24 samples in one day, and doing several times this many would be possible. We have observed on ethidium bromide stained gels that the fluorescence from the RNA deriving from this miniprep is brighter than that seen when the corresponding amount of standard-prep RNA is used. This may be due to enhanced fluorescence of RNA in the presence of ATA. However, autoradiography of the blots does not show any RNA degradation products.

This procedure would probably work fine with other methods of tissue disruption. ATA inhibits many nucleic acid binding proteins, possibly by competing for binding sites (Blumenthal and Landers 1973. BBRC 55:680-688). Therefore, the most critical factor is getting the RNA in contact with ATA before nucleases can bind to the nucleic acid and degrade it. Supported by federal grants to J.J.L. and J.C.D.

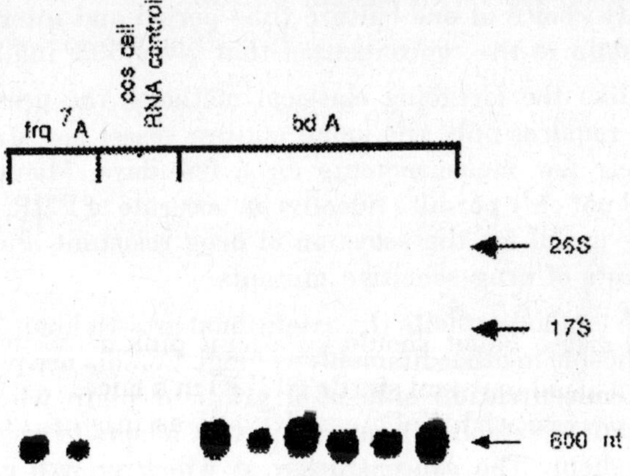

Fig. ATA mini-prep RNA probed with *ccg-1* DNA fragment. Total RNA from a series of transformants into *bd A* and *frq7 A* was examined for the presence of the *ccg-1* gene transcript (Loros et al. 1989 Science 243:385-388). Each lane contains 10 μg of RNA (1/200 of the preparation). While the fluorescence staining of the RNA extended from the 26S to below the 17S RNA bands (not shown), the hybridization revealed the presence of only a single undegraded transcript in each lane containing transformant RNA and no hybridization to the monkey cos cell RNA control.

A RAPID, EASY AND ECONOMICAL METHOD FOR ASSAY OF DRUG RESISTANCE IN NEUROSPORA

Assay of bacterial drug resistance with linear concentration diffusion gradient plates was first described by Szylbalski (1952 Science 116:46-51). We have applied the method to Neurospora.

Quantitative drug resistance may be estimated by measurement of the time-dependent increments of biomass in liquid cultures or extensional growth on race tubes with a series of concentrations at several culture time periods; however, we observe that wild type can acquire physiological resistance to many diverse drugs; therefore, it is necessary to make several measurements over periods of days to more than a week. Under those conditions, the potential maximal resistance (PMR), the least concentration that completely inhibits growth, is uncertain because the accuracy of the measurements decreases as growth increments approach zero. The relative degrees of wild-type and mutant resistance may be estimated by a judicious or arbitrary choice of one culture time period and interpolation of the data to the concentration that gives 50% inhibition.

Unlike the foregoing classical methods, the present one ideally requires only one small culture vessel per strain and relatively few measurements for a few days. Moreover the method not only permits (ideally) an accurate of PMR, but also may be useful for the selection of drug resistant mutants or revertants of drug sensitive mutants.

The method exploits the extensional growth habit and may be applicable to other filamentous fungi. Conidia are put on the "zero" concentration side of a gradient plate where they germinate to establish a mycelium that adapts to grow across the gradient. The concentration at which growth ceases is defined as the PMR. Thus, the procedure differs fundamentally from classical ones where conidia are immediately challenged with a bolus of drug. Under the latter condition, the drug may be uniquely lethal or toxic to conidial germination and/or resistance adaptation; hence the final results may not necessarily be related to the mycelial PMR.

Vogel's minimal medium with 2% each of sucrose and agar is sterilized by autoclaving and cooled to 55°C in a water bath before the addition of a drug. In preliminary experiments, several media, differing in drug concentrations by factors of 10, are prepared to determine the appropriate gradient concentration range for measurement of wild type's resistance.

Sterile plastic 5 cm diameter Petri dishes are tilted on a flat table in a sterile transfer hood by setting them on the edge of their lids. 5 ml of drug medium are put in a plate. The upper edge of the plate is marked with a felt pen to record the center of the lower end of the gradient. After the medium has solidified, the plate is set horizontally. 5 ml of drug free medium are added and allowed to solidify. The plates are inverted, set for a day to insure establishment of the diffusion gradient, inoculated with about 105 conidia in 20 μl of water in a narrow streak along the lower side of the gradient at the dish's edge and incubated at 35°C in a humid chamber.

The distance from the edge of the plate to the mycelial frontier is measured on and about the diagonal across the center. With an appropriate gradient, strains with various degrees of resistance cease to grow after two to six days. At that time, the distance to the front is a measure of PMR. The mycelial front in the central portion of the plate is usually quite uniform; hence estimate of PMR has an error variance of 10% or less.

Plots of growth distance at time intervals before PMRs are reached are curvilinear and resemble second-order decay kinetics. The transient degrees of resistance of various strains appear to be related to their PMRs in a regular fashion; however, we have not attempted to analyze the mathematical relationship.

By incorporating a dye in the medium (such as methylene blue) and photometric scans of the plates, exact measurements of the gradient linearity and its time-dependent change by lateral diffusion can be obtained (Szylbalski, pers. comm.). We have not conducted such measurements because we were only interested in the relative and not the absolute potential maximal resistances of wild type and mutants.

Age-1 and *age+* mutants are respectively inferior and superior to their wild-type parent in constitutive activities of at least 12 antioxyenzymes (Munkres et al. 1984 Mech. Age Dev. 24:83-100; Munkres 1990 Free Radical Bio. Med., in press). The accepted theory that those enzymes provide defense against free radicals and peroxides was confirmed. The PMRs of the strains with the oxidants paraquat, tert-butyl hydroperoxide and H2O2 were highly correlated with their enzyme activities (unpublished). (The theory of paraquat toxicity proposes that it generates toxic superoxide radicals in an enzyme catalyzed redox reaction.)

We conclude with technical observations related to the oxidants. On paraquat gradients, once growth ceases, usually no growth occurs for at least a week; however, on rare occasions hyphal strands grew ahead of the frontier and subsequently grew sideways. Those cells may be genetically-resistant mutants. In that context, it is probably significant paraquat has been reported to be a bacterial mutagen.

On a 0-50 mM H2O2 gradient, all strains quickly grew to the end; however on a 0-500 mM gradient, the PMRs were much less than 50 mM. Cells from the distal ends of the 0-50 mM gradients were transferred to 50-500 mM gradients whereupon the pMRs were 100-300 mM after 6-9 days. Those results apparently indicate that a 0-500 mM gradient is too steep to permit full expression of PMR without pre-adaptation on a 0-50 mM gradient. In contrast, the PMRs on 0-3 and 0-10 mM paraquat gradients were not significantly different.

On tert-butyl hydroperoxide, the PMRs of the strains were also significantly different; however, the ranges of PMRs differed on 0-1 and 0-10 mM gradients. In this case, the gradients are probably not linear. This peroxide is volatile at 35°C and its characteristic odor evolved from the plates for several days.

In conclusion, although there are minor problems in the interpretation of the assay results, such problems are probably also associated with classical assay methods. The rapidity, ease and economy of the method justify its use.

HISTOCHEMICAL DETECTION OF SUPEROXIDE RADICALS AND HYDROGEN PEROXIDE BY *AGE-1* MUTANTS OF NEUROSPORA

The *Age-1* gene family consists of 16 functionally redundant, closely linked loci on linkage group I. (Munkres 1985 Superoxide Dismutases, Vol. III, CRC Press, Boca Raton, FL). Dominant mutations at those loci reduce conidial lifespan (loc. cit.) and cause multiple deficiencies of at least 12 antioxidant enzymes; the endocellular superoxide dismutases (SOD), and exocellular cell wall bound SOD isozyme, catalase, peroxidases and reductases (Munkres et al. 1984 Mech. Age. Dev. 24:83-100; Munkres 1990 Free Radical Bio. Med., in press). Colonies of the mutants secrete superoxide radicals and H_2O_2, apparently as a consequence of their antioxidant enzyme deficiencies. The purpose of this note is to describe histochemical methods that detect the secretions.

Yellow, water-soluble nitroblue tetrazolium (NBT) is reduced by superoxide radicals to blue, water- insoluble formazan (Halliwell and Gutteridge 1985 Free Radicals in Biology and Medicine, Clarendon Press, Oxford) Although other substances may also reduce NBT, specificity of the reaction can be demonstrated by the inhibitory effect of added SOD.

Superoxide radicals dismutate to H2O2 either spontaneously or by SOD catalysis. The secretion of H2O2 is detected by a colorimetric reaction with a dye that is catalyzed by an excess of peroxidase. Cultural procedures were described (Munkres and Furtek 1984 Meth. Enzymol. 105:263-270). The density of viable conidia plated should be adjusted to obtain not more than 50-100 colonies per plate. The growth and reaction of crowded colonies is retarded. In the case of wild type, it is usually preferable to test 3-4 day old colonies before conidiophores are formed. *Age-1* mutant colonies do not form conidiophores. Mutant and wild type colonies as old as 10 days can be used, but wild type conidiophores must be removed with a forcible stream of water because they obscure the reactions.

The staining solution for superoxide contains 5 mM 3-(N-morpholino) propane sulfonate-NaOH buffer, pH 7.6, and 2.5 mM of NBT. The solution is shielded from light because the dye slowly photo-oxidizes. The colonies are flooded with 5-10 ml of the solution and incubated at 35°C for 15-60 min. The solution is discarded and plates are incubated for 0.5 to 3 h at 35°C in the dark until the desired differential intensity of stain of mutant and wild type colonies is obtained. Overnight incubation increases the overall intensity, but does not increase the differential. Commercial bovine red blood cell SOD (30-300 units/ml) in the staining solution inhibits the differential reaction. Mutant colonies are intensely blue to purple, but wild type is colorless or faintly blue. Microscopically, large blue formazan granules occur on the surface of mutant colonies, whereas the faint stain of wild type is diffuse within hyphae. (In wild type, the reducing substance in unknown.)

To test hydrogen peroxide secretion, a solution of 100 mM potassium phosphate buffer, pH 6.9, 2.5 mM diaminobenzidine tetrachloride, and 5 purpurogallin units/ml of horseradish peroxidase (Type VI, Sigma Chem. Co., St. Louis, MO) is freshly prepared and shielded from light to prevent spontaneous photooxidation. Plates are flooded with sufficient solution to submerge the colonies and incubated for an hour at 35°C in the dark. After decanting the solution, plates are inverted and incubated at 35°C in the dark for 24-48 h. Mutant colonies exhibit brown halos of various diameters and intensities and are uniformly brown, but wild type colonies do not stain.

Populations of colonies of our wild type (in Oak Ridge background) reproducibly exhibit about 15% spontaneous *Age-1* mutants (Munkres and Furtek 1984 Mech. Age. Dev. 25:47-62). About 15% of the colonies are also positive in the histochemical tests. Subsequent tests by present methods (*ibid*) indicated that the positively-stained colonies are indeed *Age-1* mutants. Thus, the secretion phenotype is a general feature of the mutants. We do not know whether the high spontaneous frequency of the mutants is a feature of other wild types, but one should be aware of the phenomenon in wild type control tests.

The superoxide test has been successfully applied to colonies derived from random ascospores in crosses of mutant to wild type or the linkage testers multicent and alcoy in analysis of segregation and recombination.

BIOASSAYS OF CYCLIC GMP OR VITAMIN E

Age-1 mutants of Neurospora are conditional auxotrophs responding to cGMP or vitamin E (tocopherol). Feeding those compounds phenotypically cures their short life span and abnormal colony development. Supplement of cGMP also phenotypically cures their antioxienzyme deficiencies (Munkres 1990a Free Radical Bio. Med., in press). An analogous class of

Saccharomyces mutants is doubly-deficient in two superoxide dismutase isoenzymes and, consequently, growth is inhibited by various oxidative stresses. Supplements of minute concentrations of cGMP or vitamin E permit those mutants to grow under oxidative stress (Munkres 1990b Free Radical Bio. Med.).

Dose-response analyses of the mold and yeast mutants indicate proportionality from as little as 10(-15) M to optima of 10(-12) M to 10(-9) M with signal amplifications of 10 to 20-fold. (loc. cit. and unpublished) Those observations indicate the feasibility of development of sensitive bioassays as an economical alternative to expensive chemical assays or as tools for physiological and biochemical analyses of the function of those molecules.

The *Age-1* mutants are deposited at the Fungal Genetics Stock Center. Our recent research has used *Age-1.7* extensively. Our four yeast mutant stocks are randomly-selected, single-colony isolates from four brewer's and vintner's strains. Mutant no. 9, derived from Red Star lager brewer's yeast, is exceptionally responsive to cGMP. We will provide the yeast stocks to interested parties or the original strains may be obtained from wine and beer makers supply shops.

A USEFUL GEL MEDIUM FOR NEUROSPORA

This note describes a useful gel medium for Neurospora with properties superior to agar-gel medium in several respects. An important property of the gel is cold-solubility which permits gentle removal of a mycelium from all types of culture vessels. Here, specific application to race tubes is described. Other merits of the gel medium will be noted.

A block copolymer of polyoxypropylene-polyoxyethylene is available under the trade name Pluronic\ F 127 Prill from BASF Corporation, Chem. Div., 100 Cherry Hill Road, Parsippany, NJ 07054. Free 100 g samples are available and no less than 40 lb. drums may be purchased. The polymer gels above 18°C and dissolves at 10°C or below. The manufacturer reports that the material is non-toxic to rats and rabbits and negative in the Ames Salmonella mutagen assays. Commercially, it is used in cold-water detergent cleansers.

A suspension of 20% w/v of the polymer is dissolved in Vogel's minimal medium containing 2% sucrose by stirring it magnetically overnight at 10°C. (Concentrations less than 20% form unsuitably soft gels or viscous solutions above 18°C.) The solution is sterilized by autoclaving it 15 min at 136°C. The gel is melted by setting it at 10°C for 4-8 h or overnight. (The time to melt is a function of the rate of cooling which, in turn, is a function of the solution volume.) The solution is dispensed to culture vessels at room temperature where it quickly gels.

The steady-state extensional growth rates of Oak Ridge wild type in race tubes on the gel medium at 22 or 35°C are essentially the same as those on 2% purified agar (Difco, Detroit, MI) medium. The long ribbon of mycelium may be removed from a race tube by chilling it to 10°C and decanting the contents. The yield of conidia by a wild type strain on slants of the gel was as good as that on purified agar. Comparative measurements of mycelial and conidial yields on slants of the gel medium should be feasible.

The gel provides an alternative to purified agar for nutritionally-exacting experiments. Unlike agar gels, the gel is highly translucent: a property that may be of merit for microscopic or photometric observations of the mycelium in situ. Comparing 20% synthetic and 2% agar gels, the polymer's cost is one half that of purified agar and equal to that of standard agar.

MODIFICATION BY LIGHT OF NITRITE RELEASE AND ACCUMULATION BY *NEUROSPORA CRASSA*

Activity of nitrate reductase (NR) induced by nitrate in the growth medium can be tested by diazotation of the nitrite formed during an activity test (Snell and Snell, 1949, Colorimetric Methods of Analysis, pp. 802-807). Pink instead of colorless controls (NADPH omitted) in these tests indicated nitrite present in the samples without NR activity and caused us to assay for nitrite excreted into the culture medium and for nitrite accumulated in the mycelia of *Neurospora crassa* strain *al-2;bd*.

Conidia for inoculation of the experimental cultures were washed from conidiating mycelia in darkness under sterile conditions and kept in sterile water in darkness at 5°C. For the experiments they were used between 1 and 14 days after harvest. Neurospora mycelia were grown from these conidia (1.5 x 10(8) per 300 ml in 2 l Erlenmeyer flasks) with modified Vogel's medium (25 mM NH2NO3 replaced by 50 mM NaNO3 and supplemented with 1% glucose) on a gyratory shaker at 80 rpm and 30°C for 16 h. In each experiment four independent culture flasks were kept in darkness, and four received 3W m(-2) of white light during their growth phase. The eight mycelia were harvested separately by filtration and the eight filtrates (each 300 ml medium) kept for nitrite assays. Three hundred milligrams of each washed mycelium was permeated for 30 min with a permeation solution containing 5% propanol (W. Ullrich, pers. commun.) and assayed for accumulated nitrite. NR activity was determined according to established procedures (Garrett and Nason 1969. J. Biol. Chem. 244:2870-2882) modified by Scheideler and Ninnemann 1986. Analyt. Biochem. 154:29-33.

Large amounts of nitrite were produced by mycelia grown on nitrate as sole nitrogen source: Table 1 shows the amounts of nitrite released into the medium or accumulated in the mycelia after 16 h of growth in NO3- medium and kept in darkness (D) or in light (L). The values were standardized by relating them to 1 g mycelium (fresh weight).

The absolute amounts of nitrite released and accumulated varied between the standardized experiments on various days. Our suspicion that the age of the aqueous conidial stock solution used for inoculation of the experiments might influence the extent of nitrite formation of the mycelia was not substantiated. But independent of the absolute amounts of nitrite, significantly more nitrite was always found in filtrates of dark grown mycelia than in those of irradiated ones. The amounts of nitrite accumulated in mycelia were about equal in dark and light grown Neurospora with a tendency of homogenates of each culture and showed no difference between dark and irradiated mycelia (values of 500-1000 nmoles NO2- formed/mg protein in 10 min at 30°C).

The added amounts of accumulated plus excreted nitrite ("total NO2-") of dark mycelia exceeded total nitrite of irradiated cultures (Table 1, column 3), i.e. in all, dark mycelia reduced less nitrite to ammonia than irradiated ones.

The high amounts of nitrite found in the media and especially in mycelia suggest that the rate of nitrate reduction exceeds the rate of nitrite reduction. The amounts of nitrite excreted into the medium resulted in NO2- concentrations of 150-250 μM for cultures grown for 16 h in nitrate medium. The same order of magnitude of released nitrite was reported for green algae under low CO2 tension and high irradiance (Azuara and Aparicio 1983. Plant Physiol. 71:286-290; Quinones and Aparicio, in: Inorganic Nitrogen Metabolism, W.R. Ullrich et al., eds. in press). A rough estimate of the minimal nitrite concentration in the mycelium can be based on 1 g fresh mycelium corresponding to 1.0-1.5 ml volume. If one third of it (or less) is cytoplasmic space, the detected amounts of 10-20 μmoles nitrite per g mycelium (fresh weight) equal 20-60 mM nitrite in Neurospora cells, a concentration the cells should have to compartimentize in order not to become poisoned (with 16 mM nitrite as nitrogen source, Neurospora does not grow anymore, Ninnemann, unpubl.).

The data of Table 1 conform to the hypothesis that nitrite reductase of Neurospora becomes activated by irradiation applied during culture growth resulting in lower amounts of total nitrite (accumulated plus released). A similar conclusion was reached for the system of Monoraphilium braunii, in which blue light appears to be required for the biosynthesis of nitrite reductase (Quinones and Aparicio, in: Inorganic Nitrogen Metabolism, W.R. Ullrich et al., eds. in press). Also, a change of nitrate uptake and nitrite export systems by light might be considered.

THE RIBOSOMAL REPEAT OF *ASPERGILLUS NIGER* AND ITS EFFECTS ON TRANSFORMATION FREQUENCY

The ribosomal RNA genes have been used in studies of a variety of phenomena. These include studies of molecular phylogeny, transcription, recombination and transformation. In *Saccharomyces cerevisiae,* where integrative transformation is achieved by homologous recombination, the presence of the ribosomal repeat unit in the transformation vector greatly increases the frequency of transformation (Szostak and Wu 1979. Plasmid 2:536- 554; Smolik-Utlaut and Petes 1983. Mol. Cell. Biol. 3:1204-1211). In the filamentous fungus *Aspergillus nidulans*, where integrative transformation can be achieved either via homologous or heterologous recombination, the presence of ribosomal repeat sequences in the transformation vector has no effect on the frequency of transformation (Tilburn et al. 1983. Gene 26:205-221). [Ed.: This is also the case in *Neurospora crassa*: Russell et al. 1989. BBA 1008:243-246]. Here we report the molecular cloning of the ribosomal repeat unit from *Aspergillus niger*. We have found that the presence of cloned ribosomal DNA sequences from *A. niger* increases the frequency of transformation in *A. niger*.

1. Cloning of the Ribosomal Repeat Unit of A. Niger

A genomic library of wild type *A. niger* (Kelly and Hynes 1985. EMBO J 4:475-479) DNA, partially digested with *Mbo*I, was constructed in the lambda based replacement vector EMBL 3. The library was screened with a 32P labelled plasmid probe, pMN1, containing the *A. nidulans* ribosomal repeat unit (Borsuk et al. 1982. Gene 17:147-152). Several plaques were identified that hybridized to pMN1. *Sal*I digested DNA prepared from these plaques showed a band of approximately 7.8 kb, and Southern blot analysis showed this band to have homology to pMN1.

That this band represented an entire ribosomal repeat unit was inferred from the finding from Southern blot analysis that *A. niger* DNA cut with a series of enzymes, including *Sal*I, *Bam*HI and *Bgl*II, gave a hybridizing band of approximately 7.8 kb representing one ribosomal repeat unit. This 7.8 kb sequence was cloned into the *Sal*I site of pBR322 and the recombinant plasmid was designated pANiR1. A partial restriction map was determined and is shown in Figure 1. The coding regions are also shown as determined both by hybridizing restriction fragments of pANiR1 to Northern blots of total RNA extracted from a wild type strain of *A. niger* and by hybridizing ribosomal RNA probes to Southern blots of various digests of pANiR1. The position of the genes, and of restriction sites, was compared with those found by Borsuk et al. (1982. Gene 17:147-152) and Lockington et al. (1982. Gene 20:135-137) in the cloned *A. nidulans* ribosomal repeat DNA. The position of the genes was found to be conserved, as were the restriction sites within the coding regions. The restriction sites within the non-transcribed spacer region were not conserved.

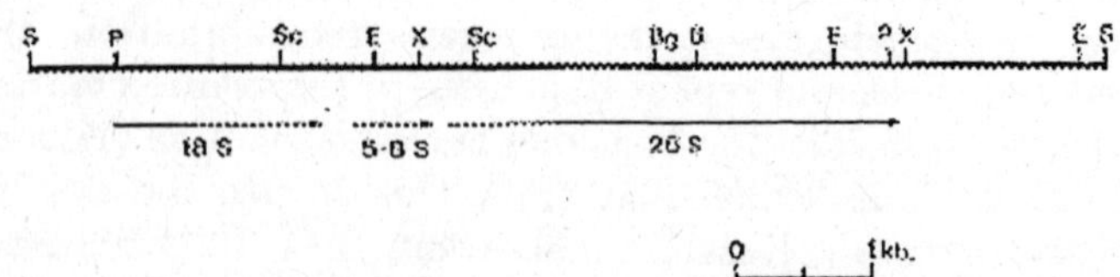

Figure. Restriction endonuclease map of the *A. niger* ribosomal repeat. Restriction endonuclease sites: S - *Sal*I; P - *Pst*I; Sc - *Sac*I; E - *Eco*RI; X - *Xho*I; Bg - *Bgl*II; B - *Bam*H1 The lines below the map indicate the positions of the rRNA genes.

The Effect of the Ribosomal Repeat unit of A. Niger of the Frequency of Transformation

Transformations of *A. niger* were performed using the *amdS* gene of *A. nidulans* as a dominant selective marker (Kelly and Hynes 1985. EMBO J 4:475-479). The plasmid used, p3SR2, contained the *amdS* gene cloned into pBR322 (Hynes et al. 1983. Mol. Cell. Biol. 3:1430-1439). Transformation frequencies were compared in experiments using p3SR2 alone, p3SR2 plus equimolar amounts of pMN1, and p3SR2 plus equimolar amounts of pANiR1 (Table I). Higher frequencies of transformation were consistently found when pANiR1 was present, but not when pMN1 was present. This result was repeatable, and did not depend on the particular *amdS* containing plasmid used. Transformation frequencies of *A. niger* using *amdS* selection were low, possibly due to the observation that transformants contain multiple copies of the *amdS* gene, and a transformant containing a single copy of the *amdS* gene may not grow sufficiently to be detected.

DNA was extracted from eight strains cotransformed with pANiR1 and p3SR2, digested with *Hin*dIII, separated by agarose gel electrophoresis, transferred to nylon membrane, and hybridized with the insert of p3SR2. The membrane was washed and rehybridized with the insert of pANiR1. There are no *Hin*dIII sites in the ribosomal repeat unit, and as expected, the insert of pANiR1 hybridized to a band of very high molecular weight, too large to quantify on standard agarose gels. There are no *Hin*dIII sites in p3SR2, and no hybridization is observed between p3SR2 and wildtype *A. niger* DNA. In the transformant DNA, the *amdS* probe also hybridized to a band of very high molecular weight. Thus it is probable that the *amdS* sequences integrated at the ribosomal repeat region.

In order to determine if this result was entirely dependent upon the selectable marker in the recipient strain, transformations were performed using the cloned *A. niger pyrG* gene (pAB4.1) as a selectable marker in a *pyrG* auxotrophic strain of *A. niger* (van Hartingsveldt et al. 1987. Molec. Gen. Genet. 206:71-75). The plasmid pAB4.1 transformed the *A. niger* strain AB4.1 at a much higher frequency than observed for p3SR2 and the wild type strain. Numbers of transformants in experiments using pAB4.1 alone were compared with numbers of transformants using both pAB4.1 and pANiR1. In four independent experiments, the transformation frequency was 3.7-, 3.9-, 4.0- and 4.5-fold higher in the presence of pANiR1 than in its absence. These experiments showed that the effect of pANiR1 on transformation frequency was neither recipient strain nor selectable marker dependent. Various subclones of pANiR1 were tested in an attempt to localize the effect. Neither the *Eco*RI-*Eco*RI fragment nor the small *Eco*RI-*Sal*I fragment had an effect on transformation frequency when present as inserts in pBR322 in cotransformation experiments. Further, the increase in transformation frequency was not apparent when the ribosomal repeat unit was inserted into the same plasmid as the *amdS* selectable marker, but this may be due to the reduction in transformation frequency due to the very large size of the plasmid masking the five fold effect.

A NEW REPORT CONCERNING NUCLEAR DNA CONTENT AND PREMEIOTIC DNA SYNTHESIS IN FUNGI

R. Duran and P. M. Gray have published results of considerable interest that may escape notice because of the title and the journal ("Nuclear DNA, an adjunct to morphology in fungal taxonomy". Mycotaxon 36:205-219, 1989). Data are presented on nuclear DNA content of species of Neurospora and smut fungi and on the time of premeiotic DNA synthesis.

What is the 1C DNA value for Neurospora Crassa?

Three estimates have previously been published for Neurospora, based on chemical extraction of microconidia (28 x 10(9) daltons, ~42 megabases. Horowitz and Macleod 1960), renaturation kinetics (18 x 10(9) daltons, ~27 megabases. Krumlauf and Marzluf 1980), and electrophoretic karyotyping (31x 10(9) daltons, ~47 megabases. Orbach et al. 1988). Duran and Gray provide yet another estimate based on microfluorometric measurements of individual nuclei stained with Schiff reagent. This method appears to have distinct advantages over absorbance microphotometry. Saccharomyces strain ATCC 26109 was used as reference standard and was assigned a 1C value of 1.05 x 10(10) daltons based on the work of others.

The value they report for Neurospora microconidia, 27 x 10(9) daltons (~40 megabases per nucleus), is close to that obtained 30 years earlier by Horowitz and Macleod. This is about 1.5 times greater, however, than the estimate of Krumlauf and Marzluf. The discrepancy might perhaps be rationalized if renaturation kinetics gave the correct basic 1C value and if the higher values from microconidia resulted because half the nuclei were post- S rather than pre-S. But the measurements of Duran and Gray suggest that the genome is unreplicated in microconidia. Histograms for individual microconidial nuclei show a unimodal rather than bimodal distribution, and the mean DNA value for mycelial nuclei is greater than that for microconidia, as expected for a population with nuclei undergoing replication. The concordance of estimates other than that based on reassociation kinetics seems to favor the 40 megabase 1C value and suggests that the value obtained using Cot curves may be too low.

When does Premeiotic DNA Synthesis Occur in Fungi?

Duran and Gray also used their microfluorometric method to determine the DNA content of nuclei after karyogamy in young asci of *N. teterasperma* and in teleospores of seven species of smut fungi. DNA appeared to be replicating still in the diploid fusion nuclei. DNA values for nuclei after karyogamy ranged from 2C to 4C in all eight species. This is contrary to accepted wisdom that premeiotic DNA synthesis has already been completed in the haploid pronuclei before karyogamy, a conclusion based on microphotometric absorbance measurements of Feulgen-stained nuclei in *Neotiella* (Rossen and Westergaard 1966 Compt. Rend. Trav. Lab. Carlsberg

35:261-386), *Sordaria fimicola* (Bell and Thierrien 1977 Can. J. Genet. Cytol. 19:359-370), *Neurospora crassa* (Iyengar et al. 1977 Genet. Res. 29:1-8), and *Schizophyllum* (Carmi et al. 1978 Genet. Res. 31:215-226). Similarly, completion of synthesis before fusion was shown by 32P incorporation in *Coprinus cinereus* (Lu and Jeng 1975 J. Cell Sci. 17:461-470) and by fluorescence of propidium iodide stained nuclei in wild-type *Coprinus macrorhizus* (Oishi et al. 1982 Arch. Microbiol. 132:372-374). In contrast, Bayman and Collins (1990 Mycologia 82:170-174), using fluorescence of mithramycin stained nuclei, have found that premeiotic DNA synthesis follows karyogamy in a homothallic isolate resembling *Coprinus patouillardii*. Postfusion synthesis was also found by Oishii et al. in a mutant of *C. macrorhizus* in which the nuclei undergoing fusion are identical.

NEW MULTICENT LINKAGE TESTERS FOR CENTROMERE-LINKED GENES AND REARRANGEMENTS IN NEUROSPORA

A tester described in 1972 contained readily-scored markers near the centromeres of all seven linkage groups (Perkins, Neurospora Newsl. 19:33). That tester, now called *multicent-1,* was somewhat more laborious to score than *alcoy* (Perkins et al. 1969 Genetica 40:247-278) or its successor *alcoy;csp-2* (Perkins and Björkman 1979 Neurospora Newsl. 26:9-10). *multicent-1* had the advantage over *alcoy* of requiring no follow-up cross to distinguish alternatives, and it was somewhat more effective in detecting linkage of left-arm markers. Because *alcoy* itself contains three translocations, *multicent-1* was more likely to identify the chromosomes involved in new centromere-linked translocations (Perkins and Barry 1977 Adv. Genet. 19:133-285).

One disadvantage with *multicent-1* was the marker *balloon,* whose restricted colonial growth meant that most of the other markers could be scored readily only in the *bal+* half of the progeny. *bal* was also inconvenient for fertilization and stock preservation. We have therefore developed three new multicent testers. These are designated *multicent-3* to *-5*. Independently, Metzenberg et al. (1984 Neurospora Newsl. 31:35-39) devised a strain designated *multicent-2,* with six centromeres marked.

This has been used for mapping cloned DNA fragments. The *multicent-3* tester incorporates two changes. *arg-5* is substituted for *bal* in linkage group II, and a long inversion, OY323, has been inserted as a crossover-suppressor in linkage group I. Progeny are isolated either to complete medium or to minimal supplemented with arginine and pyridoxine. As a result of the heterozygous inversion, mating type shows little or no recombination with markers throughout two-thirds of I, which is the longest linkage group.

In *multicent-4* also, *arg-5* is substituted for *bal*. In addition, *psi* replaces *pdx* as a marker for IV. *psi* (protein-synthesis-inhibited) is a readily-scored temperature-sensitive conditional mutant that grows normally on minimal medium at 25°C but does not grow at 34°. Ascospores must be germinated at 25° because *psi* progeny do not survive at the restrictive temperature. Replacing *pdx* with *psi* means that only two media are required. *arg-5* is scored by transferring progeny to minimal medium at 25°, *psi* by transfer to arginine-supplemented minimal at 34°.

Multicent-5 differs from *multicent-4* only in having the OY323 inversion present in linkage group I. *multicent-5* is preferred for mapping mutants that are scorable by vegetative phenotype. As with the other testers, unmapped translocations can be recognized by linkage between markers that are normally independent. If it should be desired to determine marker-breakpoint linkage by scoring for aborted ascospores in progeny tests, it is possible to score the new rearrangement only in the noninversion half of the progeny; these are recognized as noninversion because they are of the same mating type as the noninversion parent.

Scoring for the other markers is as follows: Tests for mating type are most readily accomplished on fluffy lawns in petri dishes (Perkins et al. 1989 Fungal Genet. Newsl. 36:64-66). *at* (attenuated morphology) is readily scorable on minimal (with or without supplements) at two or three days (34°C). Growth is flat on the surface, with scattered specks of conidiation. *wc-1* (white collar) is expressed most clearly at 25°C or higher. Carotenoids are absent in mycelia, though not in conidia. Germinants are incubated until conidia become orange, preferably under illumination. Scoring of *ylo-1* (yellow) improves with age, and is likely to be unreliable with young cultures, especially in combination with *wc*. Carotenoids in *ylo-1* look orange at first, then become yellow. *ylo-1* scoring at three or four days should be considered preliminary, and should be checked later. *acr-2* is scored clearly by resistance to 50 µg acriflavine/ml on solid medium.

With any of the testers, progeny are scored for markers sequentially beginning with the visible markers *at, wc-1*, and *ylo-1*. If linkage is apparent, the remaining markers need not be tested. If the unmapped mutant is unlinked to a visible marker, the markers that require transfer are then tested serially until linkage is seen. With translocations, the normally independent multicent markers are examined for linkages to one another.

The new multicent testers are heterokaryon-compatible with strains of OR background (*het-c het-d; het-e*). All three are available as heterokaryons in combination with the inactive-mating-type "helper" strain *am1 ad-3B cyh-1* (Table I). The heterokaryons are phenotypically wild-type and highly fertile. Although the homokaryotic testers can be used satisfactorily for crossing, using a heterokaryon as female parent saves labor by making it unnecessary to supplement the crossing medium. Fertility may also be improved when the heterokaryon is used.

LINKAGE MAPPING OF THE *GPDA* GENE OF *ASPERGILLUS NIDULANS*

In the last few years many genes of several Aspergillus species have been cloned and sequenced. For many of these genes mutant alleles and genetic linkage data are also available. However, for those genes for which no mutant alleles have been isolated, genetic mapping was not possible. Here we report linkage mapping of the glyceraldehyde-3- phosphate dehydrogenase gene (*gpdA*) of *A. nidulans* for which no mutant alleles have been isolated. The method used is applicable to all other cloned genes.

Transformation in Aspergillus frequently occurs by homologous recombination between host and vector sequences. Although the frequency of homologous recombination does not need to be identical for different sequences, a vector containing sequences of the gene to be mapped will often integrate at the chromosomal locus of this gene. In this way, *A. nidulans* ArgB[pAN5-41B]15 was obtained (vector pAN5-41B contains the *lacZ* gene fused to the promoter region of the *gpdA* gene of *A. nidulans* ; Van Gorcom et al. 1986 Gene 48:211-217). Southern blot analysis has shown that this strain contains a single copy of a (functional) *lacZ* gene at the *gpdA* locus. Parasexual analysis of this strain with master strain MSE (*A. nidulans* FGSC A288) was carried out. Segregation of the *lacZ* marker (integrated at the *gpdA* locus) was analysed.

Transformation of Neurospora *pyr-4* with Defective Donor Genes

Using the Vollmer/Orbach transformation protocol, transformation frequencies of a *pyr-4* (OMP decarboxylase) strain of *Neurospora crassa* of circa 10(3)/µg are routinely achievable. At these levels of transformation, it is feasible to screen out ectopic integrations and look specifically for homologous integration events. Homologous integrants were sought by transforming a *pyr-4* recipient with interrupted or incomplete copies of the cloned *pyr-4* gene derived from the *pyr+* clone in plasmid pFB6, selecting by complementation of the pyrimidine auxotrophy in the recipient.

Defective clones used as donors were of two types: 1) Clones inactivated by the insertion of the transposon Tn1000 into the *pyr-4* gene, the method of construction employing the F-mediated transfer of the *pyr-4+*, *amp*R plasmid by conjugation into a *pyr-4-*, *amp*R recipient, and identifying *pyr-4-*, *amp*R exconjugants. 2) Inactivating *pyr-4+* clones by cutting with *Bam*H1 within the distal portion of the coding region and making a translational fusion with lacZ, this construct having ß-galactosidase but not OMP decarboxylase activity.

Using two strains inactivated by Tn1000 (gd5 and gd7) and one *pyr- 4/lacZ* fusion as donor DNA, transformation of a *pyr-4* recipient was carried out. Pyrimidine-independent transformants were obtained at low frequency with each of the donor clones (see table). Control reversion frequency was circa 5 x 10(-8). With the three defective donors, restoration of an intact and functional *pyr-4* gene requires integration at the homologous chromosomal locus, with a homologous recombination event between the site of the mutant lesion in the recipient and the transposon integration site or the fusion site in the donor. This was confirmed by Southern analysis.

The locations of the three donor lesions are shown in the figure, based on restriction mapping. The lesion in the recipient is clearly proximal to the fusion site with *lacZ*, although its location within the circa 1.5 kb of the coding region proximal to that site is not yet known.

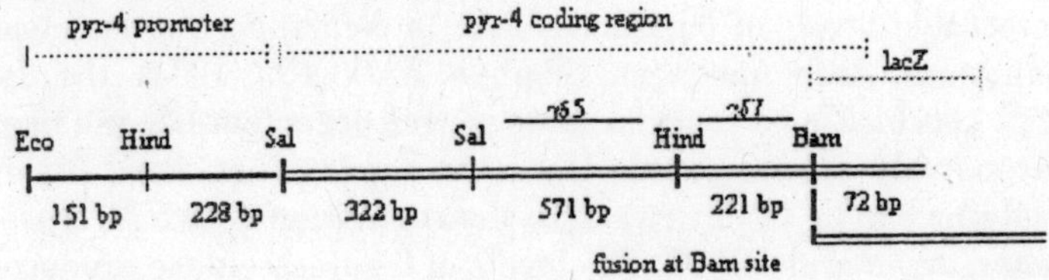

The constraints on the site of the required crossover mean that only some of the homologous integrants will give rise to pyrimidine-independent transformants. However, all such transformants are truly wild type in growth rate, unlike heterologous transformants.

wB1, a New white Mutant of *Aspergillus Nidulans* on Linkage Group VII

Several patches of white conidia were seen in green colonies of the duplication strain *proA1 pabaA6 adE20 biA1;Dp(IR IIR)yA2*, grown on 10(-2) M isonicotinic acid hydrazide; one with a mutation now designated *wB1*, has been analyzed.

A cross of the isolate to a master strain with yellow conidia and a marker on each of the eight linkage groups (MSF) yielded 683 white, 375 yellow and 337 green progeny, indicating a single gene mutation, epistatic to *yA2/yA+*. The heterokaryon with MSF bore white and yellow conidia; it gave a green diploid which, on haploidization, located *wB1* to linkage group VII.

A heterokaryon with master strain E (also with a marker on each linkage group, including *wA3* on II) had only white conidia and gave, as expected, a green diploid. White haploids from the latter, detectable as *wA3, wB1* or *wA3 wB1* by the segregation of other markers, were phenotypically indistinguishable.

Meiotic linkage was tested against only one linkage group VII marker, *wetA6*. The proportion of clearly-classifiable, non-wet progeny varied from perithecium to perithecium but the overall results suggested free, or nearly free, recombination between *wB* and *wetA*.

The use of *lacZ* gene Fusions in *Neurospora Crassa*

Systematic analyses of gene expression in diverse organisms have relied on genetic fusions in which the regulated expression of the product of the *Escherichia coli lacZ* gene, ß-galactosidase, is used to assay gene activity. Because there are low but readily detectable levels of ß-galactosidase in *Neurospora crassa* (e.g. Landman, Arch. Biochem. Biophys. 52:93-109, 1954), the use of *E. coli lacZ* as a reporter gene in this organism has not been extensively investigated. Here we report that *lacZ* fusion proteins can be used to analyze the regulation of two *N. crassa genes, arg-2* and *con-10*. The levels of ß-galactosidase produced by strains carrying the fused genes indicate that they are developmentally regulated in a manner similar to the intact genes (Davis, Microbiol. Rev. 50: 280-313, 1986; Orbach, Sachs, and Yanofsky, J. Biol. Chem. in press 1990; Roberts, Berlin, Hager and Yanofsky, Mol. Cell. Biol. 8:2411-2418, 1988; Sachs and Yanofsky, in preparation). Expression of the *arg-2/lacZ* fusion gene is high in germinating conidia and under conditions of amino acid starvation; expression is reduced by growth in arginine-containing media. Expression of the *con-10/lacZ* fusion is high in conidia and in conidiating cultures. We anticipate that the use of ß-galactosidase fusions will help extend our knowledge of how these and other Neurospora genes are regulated.

Experimental. Plasmid pUD284, derived from pAE1 (Orbach et al.,1990), contains *lacZ* fused in-frame to codon 10 of *arg-2*. Nucleotides 1479-2705 of *arg-2* were removed (by *Sty*I-digestion of pAE1) and replaced with the 3.4 kb *Sma*I-*Xba*I fragment of pC4Bgal (Thummel, Boulet and Lipshitz, Gene 74:445-456,1988) using appropriate DNA manipulations. The DNA regions containing the *arg-2* mRNA 5' and 3' ends were left intact. Plasmid pUD234 contains (i) the 1.9 kb *Kpn*I-*Bam*HI fragment of pCon10a, which includes the DNA upstream of the *con-10* mRNA start through codon 40 of the *con-10* coding sequence (Roberts et al., 1988), followed by (ii) the *Bam*HI *lacZ* cartridge from pMC1871 (Shapira, Chou, Richaud and

Casadaban Gene 25:71-82, 1983) and (iii) the *Bam*HI-*Xba*I fragment containing the RNA termination region from the *Aspergillus nidulans trpC* gene (Cullen, Leong, Wilson and Henner, Gene 57:21-26, 1987). Both pUD plasmids also contain truncated *N. crassa his-3* genes (DNA downstream of the *Hin*dIII site of pNH60, Legerton and Yanofsky, Gene 39:129-140, 1985) to select for site-specific integration in Neurospora, as well as additional sequences to maintain the plasmids in bacteria.

The fused genes were placed into the Neurospora genome by transformation of *his-3* spheroplasts (FGSC#462) and selection for his+ prototrophs on minimal medium. Prototrophic homokaryons were purified by microconidiation (see accompanying article). Analyses of genomic DNA from transformants by Southern blotting showed that the *lacZ* fusions were present as single copy genes integrated at the *his-3* locus.

The hydrolysis of o-nitrophenyl-ß-d-galactopyranoside (ONPG) was used to assay ß-galactosidase enzyme levels in clarified whole cell extracts (J. H. Miller, Experiments in Molecular Genetics, 1972, Cold Spring Harbor Laboratory, Cold Spring Harbor, New York). Germinating and mycelial cultures were grown in Vogels minimal medium/2% sucrose ± 0.4 mg/ml arginine, or in medium to induce microcyclic macroconidiation (Guignard et al., 1984), and harvested on Whatman 541 filter paper by vacuum filtration. After rinsing with water, the cells were resuspended (0.5-1 g wet weight cells/10 ml) in either ice cold HK (1 x HK buffer is 20 mM HEPES pH 7.9, 100 mM KCl, 2 mM DTT) or Z buffers containing 1 mM phenylmethylsulfonyl fluoride (added immediately prior to use from a 0.1 M stock prepared in isopropanol) by vortex mixing. Cells were broken by passage through a French press twice at 16000 psi. Clarified supernatants were obtained from the extracts following centrifugation at 4°C for 10 min at 17000 xg in a Sorvall SS34 rotor. For long term storage, glycerol was added to 20% (v/v); aliquots were quick-frozen on dry ice and stored at -70°C.

Neurospora cultures containing *arg-2*/*lacZ* fusions were also broken by sonication. Cells (0.1 g wet weight) were resuspended in one ml of cold Z buffer in a 1.5 ml eppendorf tube and were sonicated with two 10-second bursts (Heat Systems - Ultrasonics micro tip 415 attached to a model W225R sonicator; power level 4-5) with cooling between bursts. Clarified supernatants were obtained from extracts following centrifugation for 5 min at 4°C in an eppendorf microcentrifuge.

Protein levels were determined using the Bradford assay with bovine serum albumin as the reference standard. ß-galactosidase activity was measured as described (Miller, 1972). Endogenous ß-galactosidase in 74-OR23-IVA and FGSC #462 ranged from 0.5-2 units/mg protein in extracts prepared with the French press, and from 0.1-0.5 units/mg protein in extracts prepared by sonication.

Results. We constructed translational fusions of *lacZ* to *arg-2* and *con-10* and introduced the fusions into the *N. crassa* genome by transformation. Strains were analyzed for ß-galactosidase levels at different developmental stages. Strains containing the *con-10/lacZ* and the *arg-2/lacZ* fusion genes produced higher levels of ß-galactosidase than untransformed strains under all growth conditions examined. Compared to ß-galactosidase levels during mycelial growth in Vogel's medium (20 units), ß-galactosidase levels in the *con-10/lacZ* strain were 70-fold greater in conidia and were increased 12-fold when conidiation was induced in liquid culture by nitrogen limitation (Guignard et al., 1984). Levels of ß-galactosidase in the *arg-2/lacZ* strains were higher in germinating conidia than in mycelia.

Levels of ß-galactosidase were reduced at least 3.5-fold in mycelia by growth in arginine-containing medium. In other experiments in which we integrated *arg-2/lacZ* fusion genes into *arg-12s* and wild type *N. crassa* strains without targeting integration to a specific site, ß-galactosidase expression in minimal medium was greater for the *arg-12s* transformants, which are starved for arginine in minimal medium, than for wild type transformants. Expression of the *arg-2/lacZ* fusion was reduced more than 10-fold by growth of *arg-12s* transformants in arginine-supplemented medium.

CONCLUSION

lacZ is a reliable reporter gene in *N. crassa*. For the most rigorous quantification of *lacZ* activity, it may be necessary to separate *lacZ* from the endogenous ß-galactosidases, e.g., by using immunochemical or biochemical methods.

9

Isolation and Map Location

A new acetate-requiring mutant strain of *Neurospora crassa* (*ace-9*), has been isolated from the double mutant strain *cot-1;inl a*, by inositol-less death (Lester and Gross 1959. Science 129:572) in Vogel's medium supplemented with 0.3% sodium acetate and 2% sucrose. The mutant grows well on complex medium and on Vogel's medium supplemented with casamino acids, acetate, or acetate plus ethanol. Like *ace-2, ace-3* and *ace-4* (Okumura and Kuwana 1979. Japan J. Genet. 54:235-244), it shows very weak activity of pyruvate dehydrogenase compelx, but has normal activities of pyruvate carboxylase, pyruvate kinase and glucose-6-phosphate dehydrogenase.

DOMAIN INHIBITION LINKED LOW ACTIVITY OF PHE-TRNA SYNTHETASE FOR PARA-FLUOROPHENYLALANINE IN A MUTANT OF *ASPERGILLUS NIDULANS*

A number of different mechanisms for p-fluorophenylalanine (FPA) resistance have been described in *Aspergillus nidulans.* We present data on a new locus for FPA resistance and its biochemical change. Before this report, 20 FPA resistant loci conferring resistance to this amino acid analogue have been identified by us (Tiwary et al. 1987 Curr. Microbiol. 15:305-311) and two by Kinghorn and Pateman (1975 J. Gen. Microbiol. 86:174-184). These mutations have been mapped on five out of eight linkage groups of *A. nidulans,* there being no report of any mapping of FPA resistant loci on linkage groups III,IV or VII. A new class of FPA resistant mutants exhibiting a reduced level of phenylalanyl-tRNA synthetase was identified in a selection scheme using nitrate as the nitrogen source (Tiwary et al. 1987 Mol. Gen. Genet. 209:164-169). We have slightly modified the technique for the selection of analogue resistant mutants by substituting aspartic acid (25 mM) for sodium nitrate (0.6%) as the nitrogen source in the Czapek-Dox medium in the presence of 30 mg/ml FPA. A higher rate of survival was found on medium containing aspartate than nitrate.

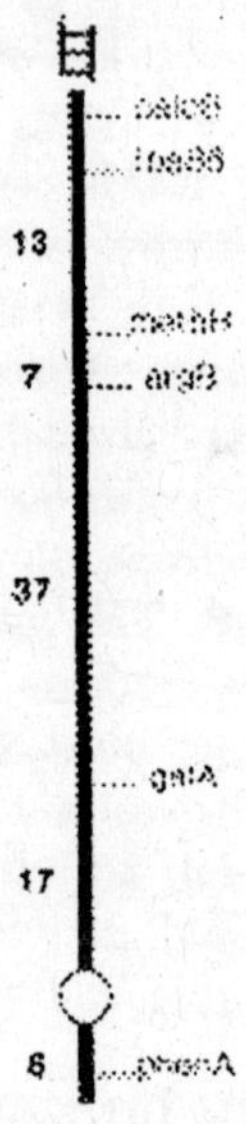

Figure. Relevant parts of linkage group III of *A. nidulans* showing the location of *fpa86* with respect to other markers

Nine mutants selected on an aspartate medium by using biA1;phenA3 (Glasgow stock no. 310) as a starter strain were found to be allelic and were given the isolation number *fpa86*. The LD50 value of FPA for the mutant was found to be 35 mg/ml whereas it was 12 mg/ml for the starter strain. Dominance test in the heterozygous diploid (*fpa86*/MSF) revealed that the new mutation was recessive to its wild-type allele. This mutation was located on linkage group III by mitotic analysis of segregants obtained after haploidization on chloral hydrate (0.02 M) of the heterozygous diploid synthesized between *fpa86* and MSF of McCully and Forbes (1965 Genet. Res. 6:352-359).

With the help of suitable crosses, *fpa86* was mapped on the left arm of linkage group III, about 13 map units left of the *methH* locus. We have also undertaken biochemical characterization of this new mutant. Table below indicates that in affinity chromatographic purification, the enzyme Phe-tRNA synthetase from the wild type as well as the mutant strain reveals similar affinity for the analogue (used as ligand) in the column. However, when the analogue (1 mM) was used as substrate in a catalytic reaction, the wild-type enzyme shows a much higher affinity for the analogue (measured in terms of ppi released) as compared to the mutant enzyme which catalyzes a very low ppi generation due to the slower rate of ATP-amino acid (analogue) exchange. This phenomenon of slow analogue activation by the mutant enzyme during the process of translation perhaps leads to its slow incorporation into proteins ultimately bringing about resistance to the analogue.

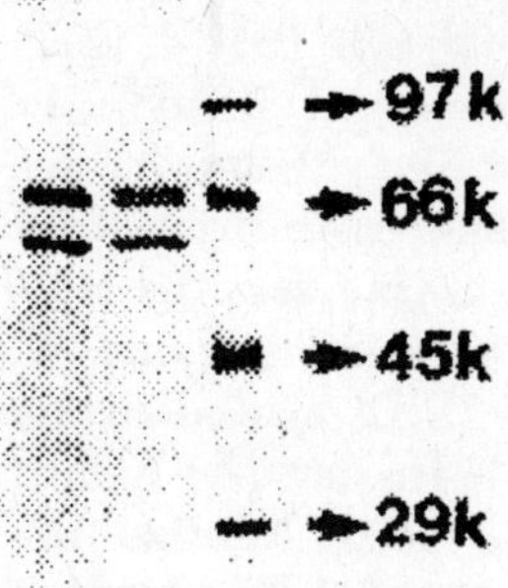

Figure. SDS-polyacrylamide gel electrophoresis (7.5%) of Phe-tRNA synthetase. (A) Wild type (B) *fpa86* (C) SDS-high molecular weight protein markers (Sigma).

Phe-tRNA synthetase of *A. nidulans* has bee reported to be a quaternary structure (alpha2ß2), the molecular mass of the alpha-subunit being 67 kdal and that of the ß-subunit 75 kdal (Rauhut et al. J. Biol. Chem. 259:6370-6375). From the results obtained, it appeared most likely that a mutation in *fpa86* caused conformational change in one of the subunits. To test the validity of this assumption, SDS-polyacrylamide gel electrophoresis (7.5%) of Phe-tRNA synthetase (Fig. 2) purified from the active fractions of the column (both wild type and the mutant) was carried out (Laemmli 1970 Nature 227:680-685). Two bands corresponding to 66 and 59 kdal were clearly visible in the case of the wild type enzyme, whereas the mutant synthetase showed the appearance of an additional polypeptide band migrating slightly faster than the ß-subunit.

This new band could be considered as a modified ß- subunit which results in functional inactivity of the enzyme in the mutant strain. However, we have no positive explanation for the discrepancy in observed sizes of the two bands if we compare our result with the earlier report on the molecular mass of the alpha and the ß subunits. Moreover, whether modification in the ß subunit occurs during activation or loading onto tRNAphe of the analogue by Phe-tRNA synthetase is yet to be determined. Experiments to investigate this issue are in progress.

Chemical modification of Phe-tRNA synthetase in FPA resistant mutants leading to alteration in the binding sites and reduced ability to bind FPA has been reported in *E. coli* (Fangman and Neidhardt 1967 J. Biol. Chem. 239:1839-1843) and *A. nidulans* (Tiwary et al. 1987 Mol. Gen. Genet. 209:167-169). It seems most likely that as a consequence of modification in the ß-subunit, there occurs a drastic decrease in the turnover number of the enzyme for the analogue due to lack of anticooperativity between domains. However, in the column the affinity of Phe-tRNA synthetase for the analogue appears to be independent of its turnover number. Furthermore, it does not really appear to matter whether all four sites in the enzyme are active or only one of them is so, because in either case the mutant enzyme is expected to show a similar affinity for the analogue to that of the wild-type enzyme in the column.

A LOTUS 1-2-3 MACRO SYSTEM TO ANALYZE A CROSS OF HAPLOID INDIVIDUALS

The use of RFLPs to construct saturated genetic maps results in the generation of very large sets of data. The management of these data, and the calculations needed to test for linkage relationships, is greatly facilitated by the use of an electronic spreadsheet on a personal computer. Using the spreadsheet Lotus 1-2-3TM (Lotus Development Corp., Cambridge, MA), we have found that the raw data can very efficiently be entered, viewed, edited, sorted, searched, and printed. We have written a program, utilizing Lotus 1-2-3 macro subcommands, that analyzes segregation data from a cross of haploid individuals in several ways.

The occurrence of phenotypes across progeny for each marker is tested for randomness by generating a chi-square statistic measuring deviation from a 1:1 ratio of the two parental phenotypes for each marker.

Markers are tested for linkage in all possible combinations by generating a chi-square statistic measuring the degree of departure from a random assortment. An example of the output generated from this procedure is shown in Table 1. Upper and lower 95% confidence limits of the recombination fraction are calculated according to the formulae of Mason, Gunst and Hess (1989, Statistical design and analysis of experiments with applications to engineering and science. Wiley Press, NY. pp. 417-419). These formulae differ slightly from those used by Bronson et al. in their program (1989, Fung. Genet. Newslet. 36:41-42).

A computation of double crossover frequencies is also available for any number of markers the operator selects (e.g. a linkage group). All possible three-way comparisons are made within the selected group. The output from this procedure includes the names of the markers and their associated single and double crossover frequencies, the most likely gene order (or orders in the event of equal probabilities), and an estimate of interference.

All output is written into the spreadsheet, thus allowing examination with the Lotus searching features as Reeves (Biotechniques 6:12-14, 1988) has suggested. Printouts of all or selected data are available with the Lotus print commands. The program was written in Lotus 1-2-3 version 2.1 and will run on any computer which supports this software. The program is available from either author - please provide a 5.25" diskette. Computational details and printed copies of the program are available from the first author.

This work was supported by Rockefeller Foundation Grant GA AS 8630 to SAL and the USDA. Mention of a trademark, proprietary product, or vendor anywhere in this paper does not constitute a guarantee or warranty of the product by the USDA/

ARS and does not imply its approval to the exclusion of other products or vendors that also may be suitable.

FAST AND RELIABLE MINI-PREP RNA EXTRACTION FROM *NEUROSPORA CRASSA*

We have developed a method for isolating high quality total RNA from *N. crassa* mycelia that reliably yields large quantities. It is possible to extract more than 50 minipreps at once.

Procedures of RNA extraction published so far follow roughly three different approaches, where phenol/chloroform, guanidinium salts, and/or LiCl is used. However, these protocols have some disadvantages. They are either time consuming, or yield low amounts of RNA, or are not powerful enough for cultures containing high levels of nucleases. Extraction from starved cultures or plate mycelia with the LiCl method (Chambers and Russo, Fungal Genet. Newsl. 1987. 33:22-24) gave completely degraded RNA as judged by gel electrophoresis. In trying to improve the method we modified the protocol used for Chlamydomonas by Gromoff et al. (Mol. Cell. Biol. 1989. 9:3911-3918) who combined the advantages of phenol extraction and LiCl precipitation. The procedure is suitable for rapidly grown mycelia, starved mycelia, mycelia from plates (4 days old) and single colonies of *N. crassa* grown on a sorbose plate. With this protocol, from 20 to 300 mg (dry weight) of mycelium, grown in liquid media or on plates, yielded between 200 μg and 3 mg of RNA depending on the growth conditions and strains. From a single colony we could isolate up to 5 μg total RNA.

Mycelium from a liquid or surface grown culture was harvested, washed, dried thoroughly with filter paper and frozen in liquid nitrogen. Single colonies including agar were stamped out using an inverted Pasteur pipette and frozen.

The procedure of extraction given below is described for Eppendorf tubes but it is also possible to scale up the volumes. All manipulations were performed at room temperature if not stated otherwise.

- Pulverize the mycelium in a mortar with liquid nitrogen
- Transfer the powder into a mixture of 0.75 ml lysis buffer (0.6 M NaCl, 10 mM EDTA, 100 mM Tris HCl, pH 8.0, 4% SDS) and 0.75 ml phenol (saturated with 0.1

M Tris HCl, pH 8.0) in a 2 ml Eppendorf tube. The tube can be filled with powdered mycelium

- Shake for 15-20 min (Eppendorf Rotationsmischer 3300)
- Centrifuge for 10 min, 10,000 rpm
- Transfer the upper phase into an equal volume of phenol (saturated with 0.1 M Tris HCl, pH 8.0) and vortex
- Centrifuge for 10 min, 10,000 rpm
- Add 0.75 volumes of 8 M LiCl to the upper phase
- Store overnight at 4°C
- Vortex briefly and centrifuge for 10 min, 10,000 rpm
- Resuspend the pellet, which is not always visible, in 0.3 ml double distilled water, mix with 0.03 ml 3 M Na-acetate (pH 5.2) and 0.75 ml ethanol
- Store at -20°C for 2 h or at -70°C for 30 min
- Centrifuge for 10 min, 10,000 rpm
- Discard the supernatant and wash the precipitate with 70% ethanol
- Dry the RNA pellet and redissolve it in DEPC treated (diethyl polycarbonate) water
- Store the RNA solution at -70°C

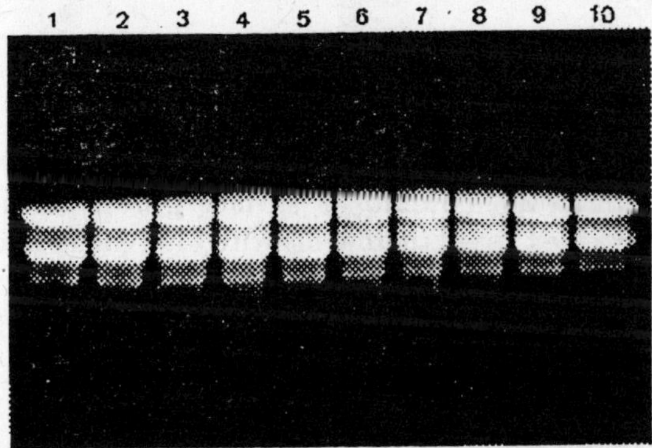

Figure. Total RNAs (10 µg per lane) from 10 different preparations (lanes 1-10) were electrophoresed on a 1.2% agarose gel containing 2% formaldehyde. Visualization was done by staining with ethidium bromide and irradiation with UV. The major bands correspond to 28S and 18S rRNA, faint bands to 23S, 16S and 5S rRNA.

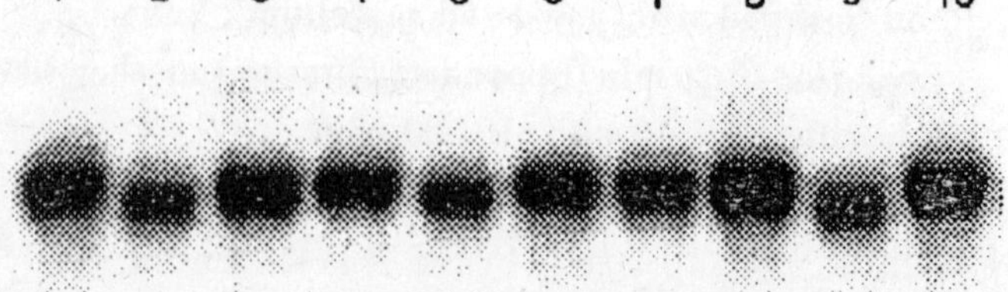

Figure. Autoradiogram after Northern blotting and hybridization to 32P-labelled cDNA insert of arbitrarily chosen clone N6 (T. Sommer et al. 1989. Nucl. Acids Research 17:5713-5723). Northern analysis was done according to R.A. Kroczek and E. Siebert 1990. Analyt. Biochem. 184:90-95.

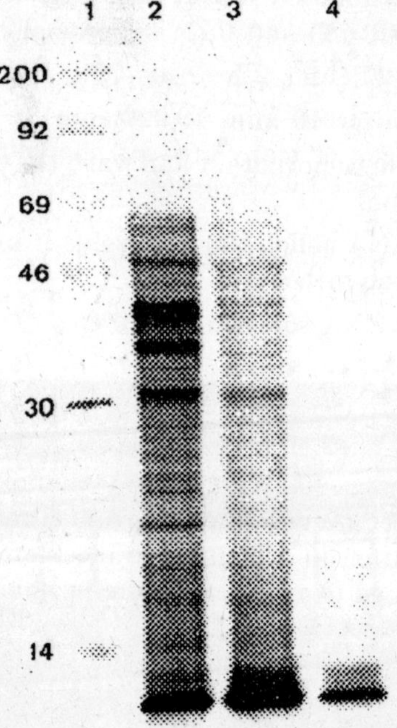

Figure. SDS-PAGE (12.5%) of 35S-methionine-labelled in vitro translation products. Lane 1: 14C-labelled marker proteins, molecular wieghts are indicated on the left (Mr x 10(-3). Lanes 2 and 3: RNA prepared from mycelia grown in rich or ammonium depleted medium respectively. RNA probes correspond to lanes 1 and 2 in Fig. 1 and 2. Lane 4: control translation without RNA.

Purity of the RNA preparations was assayed by spectrophotometric measurements. The A260/A230 and A260/A280 ratios were 2 or more, indicating the absence of any protein or polysaccharide contamination. Usually, traces of DNA present in RNA preparations can be seen in the slots of ethidium bromide stained formaldehyde gels. Based on this criterion, RNA samples prepared by the described method were free from contaminating DNA, while clear nondegraded ribosomal bands were seen in all 200 extractions we made. Northern blot analysis by hybridization with several 32P-labelled probes and in vitro translation experiments have shown clear signals indicating the presence of intact mRNA.

A MINIPREP PROCEDURE FOR ISOLATING GENOMIC DNA FROM *MAGNAPORTHE GRISEA*

We have developed a simple miniprep procedure for the isolation of genomic DNA from the ascomycete *Magnaporthe grisea*. This pathogen of many grasses, including rice, has a moderate growth rate and produces intermediate to low numbers of conidia when grown in culture. Thus, in our previous DNA preparation procedure we inoculated swirling liquid cultures with mycelium that had been fragmented in a blender rather than with conidia.

The mycelium obtained from these cultures was ground in liquid nitrogen for DNA extraction. Though the quantity and quality of DNA obtained by this method is satisfactory, the technique is too laborious for analysis of many strains. We developed the procedure described below to eliminate the need to fragment mycelium in a blender to inoculate cultures and to eliminate the need to grind mycelium in liquid nitrogen for DNA extraction. The new procedure, which relies on the enzymatic removal of cell walls and the lysis of protoplasts, should be readily adaptable to other filamentous fungi with growth characteristics similar to those of *M. grisea*.

1. Inoculate a mycelial plug into complete medium (3 g yeast extract, 3 g casamino acids, 10 g glucose/liter) in wells of tissue culture plates (24-well dishes, six columns with four rows, Corning #25820, Cell WellsTM, 16 mm diameter wells). Using 1.5 ml of complete medium per well, grow each strain in three wells in the dark at 24-

28°C without shaking until a mycelial mat covers the well (4-6 days for *M. grisea*).

2. Remove the mycelial mats from the wells, blot them on a towel to remove excess medium, and place all three mats in a well containing 1.2 ml of osmotically-stabilized enzyme solution [0.7 M NaCl, 3.3 mg NovoZymTM 234/ml (Novo Industrias, Bagsvaerd, Denmark)] for two hours at room temperature. Perform this step in a well in the same column as the one that was used for mycelial growth, thereby eliminating sample numbering for the enzyme step.
3. Pipette the digested mycelium to a microfuge tube. (The mycelial mat disintegrates with pipetting.) Pellet the protoplasts for two minutes in a microfuge at full speed. Decant the supernatant.
4. Resuspend the pellet in 600 μl of lysis buffer (50 mM Tris.HCl, pH 7.5, 100 mM EDTA, 0.5% SDS, 0.3 M sodium acetate). Heat the lysed protoplasts at 65°C for 30 minutes.
5. Extract the lysed protoplasts with 600 μl of phenol equilibrated with TE (10 mM Tris.HCl, pH 8.1, 1 mM EDTA).
6. Remove the aqueous phase to clean microfuge tubes containing 500 μl of isopropanol. Mix several times by inversion and spin immediately for 5 minutes in a microfuge.
7. Wash the pellet in 70% EtOH and air dry.
8. Dissolve the nucleic acid pellet in 100 μl TE containing 1 μg RNAse A/ml (Sigma, St. Louis).

We have successfully used this procedure to analyze several hundred transformants. One person can easily complete fifty minipreps in one day. Ten microliters of the DNA obtained (1/10 the preparation) cuts readily in 25 μl reaction mixtures with all enzymes tested (including *Bam*HI, *Eco*RI, *Hind*III, *Pst*I, *Sst*I and *Xho*I). This quantity provides a strong signal on Southern blots when 4 mm gel lanes are used. The small variation in the quantity of DNA obtained from different samples allows preliminary screening without DNA quantification.

TOWARDS A REAPPRAISAL OF THE PHENOTYPE OF THE CELL WALL DEFICIENT *FZ;SG;OS-1* ("SLIME") TRIPLE MUTANT OF *NEUROSPORA CRASSA*

Morphological mutants represent roughly 23% of seven hundred-odd distinct chromosomal loci of *N. crassa*, as listed by Perkins et al. (1982, Microbiol. Rev. 46:426). Probably the most radical phenotype among these strains is that of the *fz;sg;os-1* ("slime") triple mutant, which was isolated by Sterling Emerson (1963, Genetica 34:162) in a mutagenic experiment using an *os-1* strain. The "slime" strain has been systematically referred to in the literature as "a strain lacking cell wall and growing as protoplasts or plasmodium" (Perkins et al. 1982). Through the years, the fragile "slime" structures were frequently used as a source of organelles (Martinoia et al. 1979. Arch. Microbiol. 120:31), membranes (Scarborough, 1975. J. Biol. Chem. 250:1106) or for the study of membrane-bound enzymes (Brooks et al. 1983. J. Biol. Chem. 258:13909). "Slime" spheroplasts practically never revert to hyphal morphology; thus, the causes for impaired cell wall synthesis were investigated and attributed either to the lack of glucan synthase activity (Leal-Morales and Ruiz-Herrera, 1985. Exp. Mycol 9:28) or to improper ultrastructural characteristics of the organelles responsible for chitin synthesis: the chitosomes (Martinez et al. 1989. Biochem. Biophys. Acta 990:45).

"Slime" strains have been used in our laboratory for studying the effects of cell wall deficiency on the production of exported protein. Aside from an excess of protein secretion already reported by others (e.g. Bigger et al. 1972. J. Gen. Microbiol. 71:159), it was also observed that "slime" spheroplasts exhibit a pleiotropic defect for regulation of synthesis of several carbon-controlled catabolic exoenzymes including invertase; ß-glucosidase; alkaline protease; (Pietro et al. 1989. J. Gen. Microbiol. 135:1375); carboxy- methylcellulase; xylanase and pectinase (Polizeli et al. 1990. J. Gen. Microbiol. in press). This seems to be a general defect in the control of exoprotein production, because "slime" spheroplasts also secrete abundantly a cell wall-like particulate proteinaceous material, with self-assembling properties (Martinez et al. 1989. Arch. Microbiol. 125:25) and about fifty times more soluble protein than do wild-type strains. Moreover, the SDS-PAGE pattern of soluble extracellular protein secreted by "slime" is almost unaffected by the nutrient composition of the medium, in contrast to that of wild-type strains (unpublished).

Would these regulatory defects for protein secretion in "slime" be related to impaired cell well synthesis? This was a capital question bearing important implications concerning the role of the cell surface for processing environmental regulatory signals. This question was partially answered by reconsidering some of the details given in the original description of the "slime" strain (Emerson, 1963); specifically: that *fz;sg;os-1* ascospore segregants from crosses of "slime" and wild type ("slime-like") invariably develop into mycelium-forming cultures of abnormal morphology. These mycelial forms of "slime" (mycelial intermediate; Pietro et al. 1990. J. Gen. Microbiol. 136:121), when submitted to a protocol of filtration-enrichment selection (Nelson et al. 1975. Neurospora Newsl. 22:15), erratically produce clones with a permanent defect in cell wall biosynthesis, originating the well-known spheroplast-forming variant, stable "slime". A comparative study of mycelium-forming and spheroplast-forming derivatives of a single *fz;sg;os-1* segregant demonstrate that all defects in regulation of synthesis of exoenzymes arise concomitantly with the loss of cell wall. No new mutations seem to be selected by the procedure applied to mycelial forms for obtaining stable "slime" (Pietro et al. 1990). An analogous *N. crassa* system showing reversible formation and regeneration of stable spheroplasts was earlier reported by Selitrennikoff et al. (1981. Exp. Mycol. 5:155), using a temperature-sensitive osmotic mutant (*os-1*t, NM233t). Temperature shifts (25°C - 37°C) induce in the mutant the gain or loss, respectively, of cell wall forming activity. According to the authors, the mutant lost cell wall by incubation at 37°C.

Interestingly, the growth of "slime" spheroplasts is not restricted by several conditions which otherwise would inhibit growth of wall-forming strains, for instance the antibiotic Polyoxin B (a chitin synthase inhibitor, Selitrennikoff and Zucker, 1982. Exp. Mycol. 6:65); sorbose (a glucan synthase inhibitor; Quigley and Selitrennikoff, 1984. Exp. Mycol. 8:320); high osmotic concentrations (restrictive for growth of *os-1* mutants; Pietro et al. 1990) or temperature above 32°C (for "slimes" carrying the *cot-1* mutation; unpublished). Altogether these data stress the critical role of the cell surface for processing environmental information affecting cell growth.

In retrospect, one may realize that the real phenotype of the *fz;sg;os-1* multiple mutant strain has frequently been confused with that of its wall-less form which is, after all, just the outcome of an obscure phenomenon of vegetative transformation. It may be possible that the immediate product of Emerson's mutagenic experiment was not the wall-loss form, but a mycelium-forming strain, phenotypically similar to the mycelial "slime"-like ascospore segregants. These strains are characterized by a slow-growing, morphologically abnormal mycelium with parenchymatous surface growth, aerial hyphae practically absent and exhibiting a brown-yellowish (buff) pigmentation (Emerson, 1963). Newly isolated cultures are aconidial, but start to produce abundant microconidia- like structures after several cycles of transfer on agar medium of normal composition; otherwise the mycelial phenotype of "slime" is rather stable in standard medium (unpublished). The mycelium of "slime" is osmotically fragile and produces blobs of exudate on the surface of cultures made in solid medium, while cell wall ghosts always appear even in actively growing liquid cultures (Pietro et al. 1990).

These strains are female-sterile, a characteristic attributed to the sg mutation (Emerson, 1963) but they are very effective as fertilizing parents. Ascospores carrying the *sg* mutation germinate spontaneously, not requiring heat activation; however, if submitted to a heat treatment (45 min at 60°C) these ascospores do lose viability (unpublished). This fact suggests that the thermotolerance inherent in the ascospore is independent of the mechanism which blocks germination unless thermal or chemical stress is imposed. On cultivation in liquid medium of high osmolarity (e.g. 1.0 M sorbitol) and in the presence of 1% sorbose, mycelial forms of "slime" tend to release spheroplasts; nonetheless these spheroplasts generally revert to mycelial forms in media of normal osmolarity. Occasionally, clones with a weakened ability for synthesizing cell wall are produced (spheroplast/hyphal intermediate; Pietro et al. 1990). These forms grow as a pure spheroplast population even in media of moderate osmolarity (0.2 M sorbitol), but produce hyphae under normal culture conditions. After a few more transfers under selective conditions, spheroplast/hyphal intermediates may eventually produce stable "slime" cultures.

Apparently, during continuous growth under selective pressure fz;sg;os-1 mutants suffer a progressive and irreversible loss of cell wall synthesizing ability, reminiscent of a phenomenon of epigenetic transformation, because no new mutation can be demonstrated. Together with the loss of cell wall, other important mechanisms for recognition of environmental signals are lost as well (Selitrennikoff, 1979. Exp. Mycol. 3:363). A possible explanation for the loss of cell wall is that in *fz;sg;os-1* mutants cell growth and cell wall biosynthesis are uncoupled, the latter being too limited to encompass the growth of cells proliferating at a fast rate.

This phenomenon may be observed at early stages of ascospore germination or at the border of older colonies growing in solid medium. This effect can be dramatically exaggerated when cell wall synthesis is affected by the use of inhibitory substances such as sorbose. The loss of cell wall-synthesizing ability is gradual and irreversible, suggesting the existence of discrete units which might act as "focuses" or "primers" for cell wall synthesis, and that might be somewhat independent of immediate gene expression. These hypothetical units may be progressively lost during vegetative propagation under selective pressure, eventually giving rise to stable "slime" clones.

If the mechanism by which stable "slime" clones arise from strains with cell wall forming ability is a fascinating subject for investigation, no less intriguing is the process by which *fz;sg;os-1* ascospore segregants from crosses of spheroplasts and wild type reacquire the ability to synthesize cell wall. It seems unlikely that this phenomenon is produced by genome rearrangements. On the other hand, the female sterility of "slime" precludes genetic analysis for non-chromosomal inheritance. In this context it is worth noting that stable "slimes" cannot be recovered from "slime"/wild type heterokaryons just by conidial plating, rather it is necessary to use the time consuming method of filtration- enrichment selection. Preliminary evidence in our laboratory suggests that *fz;sg;os-1* homokaryons newly resolved from "slime"/wildtype heterokaryons may in fact be mycelial strains (unpublished).

The study of mycelial phenotypes of the "slime" strain may open new and exciting prospects for a better understanding of cell wall inheritance and morphogenesis, and of the role of the cell surface as a primary site for recognition of environmental stimuli. Other interesting aspects of the mycelial-spheroplast transformation of "slime" strains are the changes in structure and composition of the cell wall which must occur during this process. Since it is obvious that, in principle, *fz;sg;os-1* mutant strains are capable of cell wall synthesis, the causes for the reported lack of activity of polysaccharide synthesizing enzymes in "slime" spheroplasts needs reconsideration. Moreover, since it is unknown whether defective cell wall of "slime" is a result of reduced biosynthetic ability or excessive degradation, the activity of wall lytic enzymes may also deserve to be considered.

GEOGRAPHICAL DISTRIBUTION OF RESISTANCE TO SPORE KILLER IN *N. CRASSA* AND *N. INTERMEDIA*

Spore killer-2 (*Sk-2*K) is a rare meiotic drive factor found only in four (among about 2,500) collected cultures of *N. intermedia,* two from Borneo, and one each from Java and Papua New Guinea (Turner et al 1987 Fungal Genet. Newsl. 34:59-62). When strains carrying *Sk-2*K are crossed to other *N. intermedia* strains, each ascus contains four viable ascospores, which carry *Sk-2*K, and four aborted ascospores. In some populations of *N. intermedia,* one third to one half of the strains collected carry a gene, tightly linked or allelic to *Sk-2*K, which confers full or partial resistance to killing. In crosses between *Sk-2*K and a resistant (r(*Sk-2*)) strain, each ascus contains four viable *Sk-2*K progeny and four (sometimes fewer) viable r(*Sk-2*) progeny. *N. intermedia* is found around the world at all longitudes sampled, but resistance to *Sk-2*K is found only in approximately half the globe, roughly centered in the region where *Sk-2*K was found. It extends from India on the west, across the Pacific (including Japan and Australia) to Ponape. The collections that do not contain resistance include Hawaii (123 strains tested), southern Brazil (64), southeastern U.S.A. (153), Puerto Rico (84), and western Africa (48).

*Sk-2*K has never been found in *N. crassa* (among 450 tested), but fertile hybrid progeny can be produced between the two species in the laboratory. *Sk-2*K was introgressed into *N. crassa* for study because of the lack of genetic knowledge about *N. intermedia*. Resistance to killing by *Sk-2*K occurs as a minority phenotype in populations of *N. crassa*. Three points were addressed in studying the resisistant *N. crassa* strains:

(1) Allelism. One *N. crassa* strain from Malaya carries a gene for resistance that recombines at a low frequency with the previously reported r(*Sk-2*)-1 from Southeastern U.S. (Turner and Perkins 1979 Genetics 93:587-606). Other resistance factors from widely separated locations have not recombined when crossed to each other.

(2) Distribution. The range of *N. crassa* extends from Yucatan and Southeastern United States, eastward through the Caribbean, northern South America, Africa and India to Thailand. Except for equatorial Africa, it has not been found in the Southern Hemisphere. Resistance has been found in every region where *N. crassa* is found. The lowest incidence was 5/107 tested from southern U.S.A. Even looking at countries where just a few samples were collected from a limited area, resistance has been found in every country with 10 or more *N. crassa* samples, and in some with fewer.

(3) Hybridization. Resistance to *Sk-2*K gives no indication of hybridization between *N. crassa* and *N. intermedia* in nature. There is no geographical correlation between resistant populations. Resistance was absent from a number of collections of *N. intermedia* in regions where resistant *N. crassa* was found (southeastern U.S.A., Puerto Rico, and western Africa). In *N. intermedia*, r(*Sk-2*) is found in a large region from Thailand eastward to Ponape where no *N. crassa* has been found. The only overlap in resistant *N. crassa* and *N. intermedia* populations includes the eastern part of the *N. crassa* range: India, Malaya, and Thailand.

The distribution of resistance and of *Sk-2*K in *N. intermedia* suggests the (actual or recent) selective pressure of *Sk-2*K in the populations where resistance is found. The distribution in *N. crassa* could mean (1) that *Sk-2*K is or was widespread (though rare) in *N. crassa*, (2) that r(*Sk-2*) genes have some selectable function that keeps them segregating in populations independent of the presence of *Sk-2*K, or (3) that introgression took place in the area of overlap, probably with the interaction of (1) or (2).

SUBSTITUTION OF PAPER FOR SUCROSE CAN REVERSE APPARENT MALE STERILITY IN NEUROSPORA

For many years we ascertained the species of newly collected Neurospora cultures by using them to fertilize standard species testers which had grown for five days on crossing medium with sucrose as the carbon source. When new cultures would not cross to any of the testers, we tried to cross them to each other, reasoning that if any pair belonged to a previously unknown new species, it would cross successfully. This led to the discovery of *N. discreta* (Perkins and Raju 1986 Exp. Mycol. 10:323-338).

For each species, optimum tester strains were either developed or sought out among those collected. While developing testers for *N. discreta*, we turned to the use of paper instead of sucrose as the carbon source in order to increase female fertility. (The use of filter paper is described by Davis and de Serres 1970 Methods Enzymology 17A: p. 131. Some laboratories routinely use filter paper for crosses, e.g. Kinsey et al. 1980 Genetics 95:305-316, either because of fertility effects or because conidiation is greatly reduced.) We use Westergaard and Mitchell synthetic crossing medium with 1.5% agar. A piece of paper (we use Whatman chromatography paper, which comes in convenient rolls) is placed in each tube. After autoclaving, each tube must be slanted individually to orient the paper approximately along the top of the slant. Some of the cultures in the *N. discreta* group that had been female sterile on medium with sucrose made protoperithecia on the paper medium and made fertile crosses with viable ascospores when another culture in the group was used as male parent.

Some isolates from nature had not crossed to any of the species testers and had not crossed among themselves on medium with sucrose. With the use of the new *N. discreta* testers on medium with paper, some of these were identified as *N. discreta*. The others were crossed among themselves, using the medium with paper. It was found that some of the cultures that did not produce protoperithecia on sucrose medium produced abundant protoperithecia on medium with paper.

From a mixed culture that produced ascospores, pure strains of each mating type were isolated and used as reference strains. Cultures that crossed to them were tentatively designated as a new species, *N. celata*, (Turner and Fairfield, Fungal Genet. Newsl. 37:46).

We assumed that the medium with paper was affecting only the female parent, particularly after observing the profusion of protoperithecia produced on paper by some strains that apparently did not produce them on sucrose; so when we retested all the strains of the putative new species using the standard testers of other species, the more easily dispensed sucrose medium was used. We have now discovered that sucrose medium does not just affect female fertility — it will not support male fertility of certain Neurospora strains, even when the female parent is of the same species and is fully capable of crossing on sucrose to strains that do not have this conditional male sterility.

In the course of developing a revised protocol for determining the species of newly collected cultures, an unidentified strain was used as a protoperithecial parent on medium with paper. Both the *N. sitophila* tester strain and the putative *N. celata* tester successfully fertilized this strain and produced fertile perithecia with abundant ascospores. Following this observation, a set of strains previously identified as *N. celata* were crossed to *N. sitophila*. Each cross was made four times: the *N. celata* parent was used as female on both sucrose and paper, and the *N. sitophila* strain was used as female on both media. The crosses on sucrose medium were sterile, regardless of which strain was used as female. In contrast, all of the crosses on paper medium resulted in fully fertile perithecia with abundant black ascospores. Thus, the strains tentatively called *N. celata* are actually a group of *N. sitophila* that are conditionally sterile on sucrose medium, even when used as male parent.

Another project to test whether use of paper instead of sucrose would enhance male fertility was carried out using isolates that had given anomalous results on medium with sucrose. They had crossed and produced defective ascospores with *N. crassa fluffy* (this result is typical of *N. intermedia*) but had shown no crossing reaction to conidiated *N. intermedia* testers. Using medium with paper, the previously used *N. intermedia* testers were fertilized with the anomalous isolates, which did cross and make black viable ascospores on the paper medium. These isolates clearly are fertile, but when they were crossed to a conidia- producing parent on sucrose medium, there was no observable mating reaction — a phenocopy of male sterility or species incompatibility. The absence of macroconidia is the most obvious characteristic that the fluffy testers, which show enhanced fertility in many different contexts, share with the female parents grown on paper medium.

In summary, the phenomenon of conditional male sterility has been found for a subgroup of *N. sitophila* and a subgroup of *N. intermedia*. We have not had time to explore possible mechanisms for this effect.

A Putative Heterothallic Species in Neurospora

Among Neurospora cultures sent to us by Daniel Le Pierres and Assienan Bernard from the vicinity of Abidjan, Ivory Coast, were two mixed cultures that made 8-spored asci when inoculated to synthetic crossing medium of Westergaard and Mitchell (1947) with filter paper as the only carbon source. Pure cultures of each mating type were obtained from conidial platings and were found to be sterile with all of the previously known species but highly fertile with some other Ivory Coast cultures (21 in addition to the first four). For all of these strains there has not been any mating reaction observed with *N. sitophila* or *N. discreta,* but it is barely possible to assign mating type from spot crosses on a lawn of *N. crassa fl,* and under optimum conditions there is sometimes formation of barren, unbeaked perithecia with *N. intermedia*. The new strains do not act as female parents on medium to which sucrose has been added.

A serious problem in designating a new species in Neurospora is created by the crossing behavior observed for *N. discreta* (R. Maheshwari, personal communication). Sometimes two strains which have already made fertile crosses to *N. discreta* testers will fail to show any mating reaction to each other. Progeny may fail to cross to their own parents. Furthermore, most *N. discreta* strains resemble the new group in not acting as female parents on medium that contains sucrose. For these reasons, several different *N. discreta* testers were used with the new strains, and reciprocal crosses were made. No reaction was observed. The reaction of the new strains with

N. intermedia also contrasts with *N. discreta*, which never produces even rudimentary perithecia with *N. intermedia*.

We conclude that these strains constitute a new species, which has been provisionally named *Neurospora celata* (Latin meaning "concealed). Strains P4312 A (FGSC 6859) and P4376 a (FGSC 6863) have been designated as species testers. A preliminary survey was made of unidentified strains from other countries. One or two *N. celata* cultures have been identified from Congo, Gabon, Malaya, Mexico, Puerto Rico, Singapore, and Rarotonga. Collections in Mexico were made by Robert Metzenberg, and in Rarotonga by Ross Beever. The two Rarotonga strains (P4090 A, FGSC 6853; P4092 a, FGSC 6854) carry a Spore killer factor to which the standard *N. celata* testers are sensitive.

NEUROSPORA IN TEACHING

The exercise described below is a portion of a laboratory period dealing with the growth of cells and cell populations in an introductory biology course for undergraduate students. It has given good results in the two or three years that it has been used at Dartmouth College.

GROWTH OF NEUROSPORA

Select from the instructor's table a Petri plate containing a nutrient agar. The surface of this agar was inoculated a few hours earlier at the point marked "X" with a small amount of mycelium of a wild type strain of the fungus Neurospora. After an adjustment period the hyphae began growing outward, giving a series of radiating mycelial strands. Hold the plate up to the light and note the area of growth with the aid of a hand lens. Select a large straight filament at the edge of the growing region. Using a grease pencil or other marking device, mark the bottom of the dish under this filament, so that you will be able to find it again. Now place the dish, still covered, upside down on the stage of your microscope, so that you can examine this growing filament by looking through the dish bottom and the layer of agar.

The method used to measure the growth rate of these cells is as follows: Inside the ocular of your microscope has been mounted a piece of film bearing 100 regularly-spaced marks. This "scale" is in the focal plane of the ocular, and thus will be superimposed on any object seen through the microscope. To measure the growth rate of the Neurospora filament, the filament must be brought into focus (low power) and the ocular (holding the scale) must be rotated so that the growing tip of the filament moves along the scale. The growth rate of the filament can then be determined in scale divisions per minute. A stage micrometer can be used to determine the size of the scale divisions in microns. Then the growth rate can be expressed in microns per minute.

After you have aligned a growing filament with the scale as described above, begin recording in the table provided in the manual the exact time, in minutes and seconds, at which the growing tip crosses each successive scale division. Continue these measurements for 10 minutes, or until the tip has grown across 10 scale divisions. (Smallest ocular micrometer division about 15 microns.) Plot these measurements on graph paper, using time on the horizontal axis and position (scale divisions) on the vertical axis. Draw a smooth curve through the points.

Questions: Is the growth rate constant? If so, calculate the average growth rate in divisions per minute and then convert to microns per minute. Enter calculations in the table.

Having measured the normal growth rate of the Neurospora filament, you are now asked to study the effect of one of a number of externally applied substances on the growth rate of this same filament. A list of the available agents will be posted on the board. (Among those substances which have been tried are sucrose and other carbon sources, sorbose, glycerol in various concentrations, urea, toxicants like dithiocarbamates, mercurials of various types, some amino acids, yeast extract, dinitrophenol, copper salts and many others.)

Make sure the filament you have been studying is marked so that it can be found again and then remove the plate from the microscope. Remove the cover and place one drop of the substance to be tested on the marked filament. Replace the cover and center the filament under the microscope as before. As before, record the time the filament crosses each division of the ocular scale. Enter these data in the table. Continue these readings for 10 minutes or until the tip grows across 10 divisions. Plot the growth of the treated filament in the same manner as you did that of the untreated one. Connect the points with a smooth curve.

Questions: Is the growth rate of the treated filament constant? If so, calculate the average growth rate in divisions per minute and microns per minute and enter calculations in the table. If the growth rate of the treated filament is not constant, describe the nature of the growth curve. How would you describe the effect of the substance you used on the growth of a Neurospora hypha? How could you prove that the effects observed were due to the applied substance and not due to the water in which it was dissolved? How could you show that the effects observed were not caused by the brief removal of the cover? If you have time, perform experiments designed to answer these questions.

SUSKIND, S. R. GROWTH OF NEUROSPORA.

The following exercise is taken from "Principles of genetics Laboratory manual" by Hartman, Suskind and Wright (Wm. C. Brown Co., 1965) and is presented here with the permission of the authors and publisher.

THE GROWTH OF FUNGI

The purpose of this experiment is twofold. First is to examine the morphology of an ascomycete, and second, to become acquainted with the concept and use of nutritional mutants. As you know, biochemical genetic studies have relied heavily on the use of the bread mold,Neurospora crassa. The widespread use of Neurospora in biochemical genetic work is based on its rapid growth and development during versatile life cycles, its well characterized genetics (particularly at meiosis), and its simple nutritional requirements. It can be readily propagated asexually, and large populations of a particular genotype can be easily obtained.

Neurospora (wild type) will grow on a medium containing inorganic salts, a carbon source such as sucrose, and one of the B vitamins, biotin. These substances constitute the "minimal" medium, and from these components, Neurospora can synthesize all of its protoplasmic constituents including vitamins, amino acids, purine, pyrimidine and fats.

It is known that mutation can often be recognized phenotypically by the loss of the capacity of the organism to carry out a particular biochemical reaction. This is frequently accompanied by the appearance of a specific nutritional requirement reflecting the effect of the "genetic block". For example, a mutant strain may require a particular vitamin, amino acid, etc. Such mutants are termed "auxotrophic" strains. In the absence of the specific growth factor required by the mutant, the mutation becomes lethal, i.e., the organism cannot grow.

In today's experiment you will set up a bioassay for vitamin B_6 (pyridoxine) using a pyridoxineless mutant, Y-2329, of Neurospora. Three points will be emphasized in this experiment. (Each pair of students will set up the experiment.)

1. The nutritional independence of a wild type of strain, 5297 (i.e., growth on minimal medium without pyridoxine).
2. The nutritional dependence of mutant Y-2329 (i.e., no growth on minimal medium without pyridoxine).
3. The quantitative growth response to varying concentrations of pyridoxine (i.e., microbiological assay for pyridoxine).

A. Each group of students will be provided with a conidiated slant of wild type strains 5297a and the pyridoxine mutant V-2329. For each slant, transfer a loop-full of conidia to a tube containing 5 ml of sterile distilled water and vigorously mix until a suspension of conidia is obtained. These will be your inocula for the growth experiment.

B. 1. 5297a experiment.

Add 20 ml of minimal medium to each of three flasks (125 ml Erlenmeyer), stopped with cotton plugs and autoclave.

2. Y-2329 experiment.

a. Add 20 ml of minimal medium to each of three flasks (125 ml Erlenmeyer), stopped with cotton plugs and autoclave.

(Minimal control)

b. Add 20 ml of minimal medium to each of 10 flasks (125 ml Erlenmeyer), using a pyridoxine stock solution (1 ug/ml), add pyridoxine concentrations ranging from 0.1-1.0 ug pyridoxine/flask. Stopper with cotton plugs and autoclave.

C. Place flasks in cold room for 15 minutes. After the flasks have cooled to room temperature, inoculate flasks from B-1 with 2 drops each of the wild strains 5297a conidial suspension. Gently swirl flask contents. The flasks from B-2a, b are similarly inoculated with the conidial suspension of the pyridoxine mutant, Y-2329. Use sterile 1.0 ml pipettes for the inoculation of the flasks.

D. Incubate the flasks at 30C for 72-96 hours. The flasks can be stored in the cold room until next lab.

E. Harvest and weigh the mycelia. The mycelia are fished out of flasks using a small spatula. Sterile technique is not necessary at this point. Squeeze the pads as dry as you can and place each pad in a numbered depression of a spot plate. Dry in the oven (100C) for 3 hours. Weigh the dried samples on an analytical balance.

F. Plot your data, i.e., a pyridoxine response curve: mg. dry weight per pad vs. pyridoxine concentration.

Materials per 2 students:

96-120 hr. slant of 5297a - 1

120 hr. slant of Y-2329 (B_6) - 1

test tube with 5 ml sterile distilled water - 2

1.0 ml pipettes - 3

125 ml Erlenmeyer flasks - 16

baskets and trays for carrying flasks

pyridoxine stock solution 1 ug/ml - 5 ml

spatula

porcelain or glass spot plate - 2

non-absorbent cotton

minimal medium (incl. 2% sucrose) - 350 ml

References: Straus 1951 Arch. Biochem. 30:2912; Wagner and Mitchell 1964 Genetics and metabolism. 2nd ed. Wiley; p.

163. Vogel and Bonner 1958 p. 1 In Ruhland (ed.) Encyclopedia of plant physiology. Springer; Fincham and Day 1963 Fungal genetics. Davis.

MEDIUM: FRIES MINIMAL MEDIUM.

Aiuto, R . Intergenic mapping by a sorbose selection technique.

The following exercise has been found useful in laboratory sections in which adequate numbers of dissecting microscopes are not available for Neurospora ascospore isolation. It is adapted from methods presented by Lavigne (1962 NN# 2:20), Smith (1962 NN# 1:16) and Frost (as described in Fincham and Day 1962 Fungal genetics, 1st ed. Blackwell, Oxford). The technique requires the use of a biochemical mutant linked to the temperature-sensitive colonial mutant cot, crossed to another linkage group IV biochemical mutant, and the use of the growth-inhibiting carbon source, sorbose.

Ascospores from a cross of pdx-1, cot x pan-1 are harvested with a sterile loop and suspended in 5 ml of sterile distilled water. Usually a single loop of ascospores is more than sufficient for mapping purposes. Four differently supplemented flasks of medium N are prepared (2% agar, 1% sorbose, 0.1% sucrose), each containing 50 ml. One ml of the ascospore suspension is heat-shocked at 60C for 30 minutes in each of the flasks of molten medium, and two plates are poured from each flask. After 60 hours of incubation at 33C, two types of colonies can be recognized: typically sorbose-inhibited colonies three tofour mm. in diameter, and colonies about 1 mm. in diameter. These are cot^+ and cot^- individuals, respectively.

The two minimal plates reveal gene order. Most of the colonies will be cot^- and represent one-half the single crossovers in region 1, while the cot^+ colonies, far fewer in number, are one-half the doubles. Colonies scored on pyridoxine-supplemented medium (0.2 mg/ml) as cot^+ are one-half the single crossovers in region 2 less the number of cot+ colonies observed on minimal medium. Colonies scored as cot^- on plates supplemented with calcium pantothenate (0.1 mg/ml) represent the reciprocal single crossover class of region 2 less the number of cot^- colonies observed on minimal medium. Completely supplemented plates, those supplemented with both pyridoxine and calcium pantothenate, furnish the total number of viable ascospores plated, and the total number in each parental class. This latter calculation is determined by the difference of the recombinant classes (counting the values obtained for singles in region 1 and doubles twice) from the total number of colonies appearing on complete medium.

Other linkage group IV crosses can be used, such as me-1 x his-5, cot. Some mutant combinations, however, e.g., pyr-l, cot, do not lend themselves to clear-cut discrimination of $cot^+$and cot^- colonies, nor do all mutant combinations respond satisfactorily at 33C. In such instances, colonies must be isolated to appropriately supplemented non-sorbose tubes, i.e., from minimal plates to minimal tubes, and scoring is then done with the tubes.

In addition to permitting a mapping exercise with Neurospora without the use of dissecting microscopes, this method introduces the student to a plating technique, and allows instructor and class to discuss the theoretical basis for mapping genes with recombinant classes incompletely identified. The exercise can be shortened by considering only gene order (using only minimal plates) or extended to a four-point cross by isolating colonies to individual tubes, allowing these isolates to grow up at 25C on non-sorbose medium, and testing in liquid medium for the fourth marker.

The following are time-tested Neurospora teaching experiments. All of the strains mentioned are, or soon will be, deposited with the Fungal Genetics Stock Center and should not be requested from the authors (NN editor).

Experiment 1. Growth tests on Neurospora mutants of the methionine-threonine pathway.

Problem: to locate the point of blocking in a series of mutants.

Solutions needed:

1. Liquid minimal medium containing sucrose (2%) and biotin (5ug per liter).

2. Same, containing 50 mg L-methionine per liter.

3. Same, containing 150 mg DL-homocysteine thiolactone per liter.

4. Same, containing 200 mg DL- + Allo-cystathionine per liter.

5. Same, containing 50 mg L-cysteine-HCl per liter.

6. Same, containing 40 mg Na thiosulfate +.5H20 per liter.

7. Same, containing 50 mg L-threonine per liter.

8. Same, containing 50 mg DL-homoserine per liter.

9. Same, containing 50 mg L-methionine + 50 mg L-threonine per liter.

Strains Needed: Each student pair will be given a set of 8 cultures labelled with isolation numbers. The cultures are listed below.

Mutant #	Locus	FGSC #	Block
80702	cys-2	125	SO_3^{2-}-----/--->$S_2O_3^{2-}$
39816	cys-10	427	$S_2O_3^{2-}$----/-- >Cysteine
26104	me-3	112	Cysteine + Homoserine -----/---> Cystathionine
H98	me-2	283	Cystathionine-----/--- > Serine + Homocysteine
38706	me-1	560	Homocysteine -----/---> Methionine
51504	hs	471	Aspartic acid -----/---> Homoserine
35423	thr-2	2	O-Phosphohomoserine -----/---> Threonine
25a	wild type	353	(any wild type can be used)

Procedure: Each strain is to be inoculated into each of the media listed and observed for growth (+or-) after 48 and 72 hours of incubation at 25C. Use 20 ml of medium in the culture flasks provided, plug with cotton, and autoclave at 15 lbs for 10 minutes. At the same time, autoclave 8 tubes containing 2 ml of distilled water to be used for making conidial suspensions for inoculation. After the tubes have cooled, make a conidial suspension of each mutant (visibly turbid) and inoculate the flask with a drop of the suspension.

Experiment 2. Demonstration of an enzyme deficiency in a Neurospora mutant.

This experiment consists of two parts: Part A is a growth test in which the phenotypic effect of a mutation that abolishes the synthesis of D-amino acid oxidase is demonstrated. Part B is a test for the enzyme in the mutant and in wild type.

Part A. 18 flasks of supplemented minimal medium will be needed. Nine of the flasks are to be inoculated with mutant #38706, blocked between homocysteine and methionine. These flasks are supplemented as indicated below.

Flask	L-methionine	Flask	D-methionine
1	0 ug/ml	6	4 ug/ml
2	4 ug/ml	7	8 ug/ml
3	8 ug/ml	8	16 ug/ml
4	16 ug/ml	9	32 ug/ml
5	32 ug/ml		

The second set of nine flasks is identical with that listed, except that all flasks contain, in addition, 10 ug of inositol per ml. These flasks are to be inoculated with the triple mutant 38706, 89601, oxD(8). This mutant carries, besides the methionine gene, a block in the synthesis of inositol, and a block in the synthesis of the D-amino acid oxidase. (The inositol requirement is irrelevant to this experiment; it is present because the oxD mutation was isolated by the inositolless-death method).

After the flasks have been autoclaved and cooled, inoculate them and incubate at 25C for 72 to 96 hours. Fish out the mycelial mats, press out the excess medium on paper towels, and determine the wet weight of each. Part B. This experiment will use a wild type strain and mutant oxD (8) which has been freed of the methionine and inositol requirements by crossing. Grow three cultures of each in 20 ml of minimal medium contained in 125 ml Erlenmeyer flasks. Grow the cultures for 5 days at 25C. Decant the medium and wash each mycelial mat individually by adding 20 to 25 ml of water to the flask, swirling, and decanting. Repeat, using phosphate buffer, 0.1 M, pH 7.2. The cultures may be stored in the deep freeze at this point if so desired. Only four of the cultures will be needed for the assay; keep the extra pair as spares.

To one wild type culture and to one mutant culture, add 5 ml of phosphate buffer containing D-methionine at a concentration of 2 mg/ml. To the other two, add buffer without substrate. Incubate the flasks at 37 for one hour with shaking. Filter the solutions through paper and determine the keto-acids by the dinitrophenylhydrazone procedure which follows:

1. Add 1 ml dinitrophenylhydrazine (0.1% in 2N HCl) to 1 ml filtrate and let stand 5 minutes.
2. Add 2 ml absolute ethanol .
3. Add 5 ml 2.5 N NaOH and shake vigorously to mix. Let stand 5 minutes.
4. Read color in the Klett, using the green filter.

Standard Curve: In place of filtrate use 0, 10, 20, 30, 40, and 50 ug of sodium pyruvate. Calculate the amount of a-keto acid produced from D-methionine by the wild type and mutant strains, assuming that pyruvate is twice as chromogenic, gram for gram, as a ketomethiobutyrate. What reaction is catalyzed by the D-amino acid oxidase? How can you explain the results obtained in Part A?

Reference: Ohnishi, Macleod and Horowitz 1962 J. Biol. Chem. 237: 138.

Experiment 3. Accumulation of imidozole compounds by a histidine-requiring mutant. Feedback control of biosynthesis.

Into a series of 12 numbered 125 ml Erlenmeyer flasks place 20 ml of minimal medium. To each set of 6 flasks add

0, 0.1, 0.3, 0.5, 1.0 and 2.0 mg, respectively, of L-histidine. Plug the flasks with cotton and sterilize for 10 minutes in the autoclave. Inoculate one set with wild type and the other with histidineless C-84 conidia. Incubate at 25 for 4 days.

Place ½ ml of medium from each flask in which growth has occurred into a series of numbered test tubes. At the some time, set up a series of standards consisting of ½ ml minimal medium plus 0, 10, 25, 50 and 75 ug, respectively, of histidine. Determine imidazole derivatives in the solution by means of the Pauly reaction, as modified by Jorpes (1932). The procedure is explained below. Remove the mycelial mat from the flasks analyzed, press out the moisture on paper towels, and determine the wet weight of each. Record the calculated imidozole concentration for each flask and the amount of imidozole accumulated per gram of mycelium.

Determination of histidine (Jorpes, loc. cit.)

Solutions needed:

1. Sample to be tested.
2. A diazonium solution prepared as follows: to 1.5 ml of a solution containing 0.9 g sulfanilic acid and 9 ml conc. HCl in 100 ml are added 1.5 ml of a 5% sodium nitrite solution. Cool on ice for 5 minutes. Then add 6 ml of the nitrite solution with shaking, cool for 6 minutes, and add water to 50 ml. The diazonium solution should be kept cold. It keeps for 24 hours. Best results are obtained with freshly prepared sodium nitrite solution.
3. 1.1% sodium carbonate solution.

Procedure: To ½ ml of solution 1, add 2 ml of solution 2. After 1-3 hours, add 3 ml of solution 3. Read in a Klett-Summerson colorimeter with a blue filter (400-465mu) 4-8 minutes after addition of solution 3.

References: Haas et al. 1952 Genetics 37: 217; Jorpes 1932 Biochem. J. 26:1507; Ames 1955 p. 357 In McElroy and Glass (ed.), Symposium on amino acid metabolism. Johns Hopkins Press, Baltimore.

Discussion: C-84 (his-1) is blocked between imidazole glycerol phosphate (IGP) and imidozole acetol phosphate (IAP) in the pathway of histidine biosynthesis. As the mutant grows,

IGP accumulates in the mycelium and soon its dephosphorylated derivative leaks out and is found in the medium. As a control, wild type (25a) is tested and no imidazole compound is found in the medium. Both mutant and wild type take up the histidine originally added.

The phenomenon of feedback control of biosynthesis is demonstrated when it is noted that the amount of imidozole accumulated per gram of mycelium decreases as the amount of added histidine increases. The end product of a pathway has inhibited an early step in its own synthesis.

Experiment 4. Tyrosinase.

The following experiments demonstrate the thermostability and electrophoretic differences between two allelic tyrosinases of Neurospora. The alleles T^L and T^{PR} were chosen for this experiment because they are easily distinguished by both tests. If it is desired to omit the electrophoresis experiment and perform only the thermostability test, then the alleles T^S and $T^L$are recommended.

Induction: Place 20 ml of ½ strength Vogel's medium containing ½% sucrose into each of ten 125 ml Erlenmeyer flasks. Plug with cotton and sterilize for 15 minutes. Inoculate 5 of the flasks with the strain 4-137 T^L and 5 with 65-1434 T^{PR}. Incubate the flasks at 25C for 3 days. Induce 3 flasks of each strain with 2 mg sterile D-phenylalanine (2 mg/ml) per flask using aseptic technique. Leave 2 flasks of each uninduced. Incubate the flasks for an additional 24 hours at 25C on the shaker. Tyrosinase is formed during this time in those flasks containing inducer.

Remove the excess medium by pressing the mycelia between paper towels. Weigh the pooled induced T^{PR} mycelia, and extract the tyrosinase by grinding with sand in a cold mortar, using 4 ml of 0.1 M sodium phosphate buffer, pH 6, per gram of moist mycelium. Do the same with the T^L mycelium, and with the uninduced controls, yielding a total of four extracts. Keep the extracts in an ice bath during waiting periods. Spin the extracts at 10,000 x g for 10 minutes or at the highest speed of the clinical centrifuge for 10 minutes, and decant the supernatant, containing the enzyme.

Assay: Add 0.02-0.1 ml crude extract to sufficient sodium phosphate buffer (0.1M, pH 6) to give a final volume of 4 ml, and equilibrate the solution in a water bath at 30C. To start the reaction, rapidly add 1 ml DL-dopa (4 mg/ml in buffer). Shake the tube and read it in the Klett, using the blue filter. Shake the tube again and replace it in the water bath. Five minutes after the first reading was taken, make a second reading on the tube. The difference between the two readings is proportional to the tyrosinase concentration. Demonstrate this proportionality by setting up a concentration series, using either one of the two preparations you have just made. Have at least 5 points, including a zero control, in the series. The readings should be linear with concentration up to about 150 Klett units/5 minutes. Above 150-200 units, the curve departs from linearity owing to the fact that oxygen becomes limiting for the reaction.

Thermostability Test: Into 4 Klett tubes place the calculated volume of T^{PR} enzyme which would give a reading of 100-150 Klett units in 5 minutes; make up the volume to 0.5 ml with buffer. Do the same with the T^{L} preparation. Cover the tubes with foil. Keep one tube of each set in an ice bath as a zero control. Set the other six in a water bath at 59C. Remove one tube of each extract after 10 minutes and chill immediately. Repeat after 20 and 30 minutes. To the chilled samples, add 3.5 ml buffer to each tube and proceed with the assay as above. Estimate the half lives of the two enzymes. (More points may be taken on the curves if desired.)

Electrophoresis: Make up 1 liter of 0.05 M sodium phosphate buffer, pH 6. Dissolve 1 gm of bovine serum albumin in 100 ml of the buffer, and pour the solution into an enamel tray. Saturate 8 numbered paper strips in the solution, allow them to drain for a few seconds, and place on the holder of the Spinco Paper Electrophoresis Cell. Mix the BSA solution with the remainder of the buffer and pour into the reservoirs of the cell . Set the strip holder in position, close the cell, and level the buffer in the reservoirs. The enzyme solutions should contain between 1500 and 3000 Klett units per ml for best results in the electrophoresis. Apply 0.01 ml of each to a paper strip with the wire applicator. (If enough cells are available, use a mixture of the two on a third strip.) Replace the tape on the cell cover, and plug in the cell. Turn on the power supply, and set it for constant current at the rate of 1.25 ma per paper strip. Allow to run for from 16 to 24 hours. Turn off the power supply, unplug the cell, remove the cover and the strip holder. Lay the holder, containing the strips, on a flat surface. Spray the strips with a solution containing either 4 mg/ml of DL-dopa or 0.8 mg/ml of epinephrin. Dopachrome (or adrenochrome) will be formed at the position of the enzyme. Use an atomizer that gives a fine spray; do not flood the strips. Measure the distance migrated.

Henningson, K. W. Neurospora class experiments.

The following Neurospora class experiments are used in a course in biochemical genetics at the Institute of Genetics, University of Copenhagen.

Experiment 1. Tetrad analysis (ascospore color).

A cross between a mutant, asco (37402)a, with colorless ascospores (requirement for lysine) and a wild type strain, 74A, is used for tetrad analysis. Since the color of the ascospore is an autonomous character, it is possible to analyze the segregation of the alternative characters of the ascospores within the intact cell. The cross between strains of opposite mating types can be performed in two different ways: if the two strains are inoculated simultaneously, each strain acts as both female and male parent. If desired, one or the other strain can be used as female, by inoculating it before-hand on the crossing medium (Westergaard and Mitchell 1947 Am. J. Botany 34:573) to induce formation of protoperithecia. Then, mating of protoperithecia can be carried out with conidia from the other strain. Either solid or liquid medium may be used. In the case of liquid substrate, filter paper is used to support hyphae and protoperithecia.

Materials:

First day: Test tube cultures of 37402a and 74A. Three test tubes containing 10 ml liquid crossing medium + 300 mg/l lysine. 0.90% NaCl . Strips of filter paper (sterile).

Second day: 2% sodium hypochlorite. Distilled water. Microscope slides and coverslips. Petri dishes, needles and forceps for dissection.

Experimental Procedure:

First Day: Harvest conidia from both strains by pipetting about 3 ml of 0.9% NaCl into the test tube culture and shaking gently. Transfer from each strain 0.5 ml of spore suspension to the crossing medium. In each of the cultures a strip of filter paper is oriented so that about 5 cm stands out of the liquid. The cultures are incubated at 25C for 3-4 weeks.

Second Day: Analysis of the tetrads. After about 3 weeks, the pyriform perithecia contain asci with maturing ascospores. The filter paper which carries most of the perithecia is removed from the test tube and placed in a petri dish containing 2% sodium hypochlorite. After 3-5 minutes, the conidia have been killed by the hypochlorite and the filter paper can be transferred to another dish and washed with distilled water. A few perithecia then are placed on a microscope slide . Under a dissecting microscope the contents of the perithecia are removed in a drop of distilled water and the perithecial wall is removed from the slide. A coverslip is placed on top and the asci are spread by a light pressure. Under the light microscope all asci containing 4 black and 4 colorless ascospores are analyzed. The number of first and second division segregation (pre- and post-reduction) tetrads is determined and the distance between the gene for ascospore color and the centromere is calculated from the formula: ½ x percentage of post-reduction tetrads.

Reference: Stadler 1956 Genetics 41:528.

Remarks: As the ascospores approach maturity, some asci containing 8 black spores may appear. If these abnormal asci are excluded from the results, the distance between the gene and the centromere is about 14%.

Preparing Cultures: The asco strain (37402)a, FGSC#405, which requires lysine, is kept on Vogel's minimal medium + lysine 300 mg/l. Wild type strain 74A is kept on Vogel's minimal medium N (Vogel 1956 Microbial Genet. Bull. 13:42). Stock cultures are maintained and tested before use as described for arginine mutants.

Experiment 2. Isolation and characterization of biochemical mutants.

A. Isolation of biochemical mutants by filtration technique.

Conidia from a wild type strain of N. crassa are irradiated with X-rays and subjected to differential germination on liquid medium. Passage of the culture through a suitable filter at various intervals removes more and more of the germinated wild type conidia until finally the culture contains only conidia which are unable to germinate on minimal medium. After germination and isolation of these residual conidia on complete medium, many show specific growth-factor requirements.

Experimental procedure:

(1) Harvest conidia from 6 cultures of wild type strains 74A by pouring glass pearls into the flasks and shaking. Then pipette 10 ml 0.9% NaCl into each flask and shake again. Filter suspension through a G-2 glass filter. Centrifuge filtrate at 3000 x g for 3-5 minutes and resuspend pellet in 0.9% NaCl . Repeat 3 times. Determine the conidial concentration of the suspension by counting the spores in a haemocytometer. Adjust the final concentration to 1 x 10^8 spores/ml . A minimum of 10 ml of the final concentration is needed.

(2) Pipette 3 ml of the conidial suspension into each of 2 alcohol-sterilized plexiglass dishes that will fit into the X-ray machine. Irradiate one dish at 4 x 10^4 r (100 kv, 8 cm. distance, 1.7 mm aluminum filter) for 15 minutes. The second dish is kept as a non-irradiated control .

(3) Transfer the irradiated suspension to a test tube. Rinse the plexiglass dish with 3 ml 0.9% NaCl and add this to the test tube. Pipette 4 ml of this irradiated suspension into a 2-liter flask containing 400 ml Vogel's N-minimal medium + 0.5% sucrose. Prepare a control flask of the non-irradiated sample in the same manner. Place the flasks on a mechanical shaker and shake at low speed at 32C. Record inoculation time as zero time for growth.

(4) To determine the survival and germination percentages, dilute 1 ml of the remaining irradiated and control suspensions to 1,000 and 100 conidia/ml. Transfer 5 samples of 1 ml from each dilution to 5 tubes each containing 15 ml of complete medium with sorbose (melted and kept at 42C). Shake and pour into sterile petri plates. Incubate at 20-25C for 2-3 days (may be stored at 4C).

(5) To remove the germinated wild type spores, which have formed hyphae, filter the contents of the two flasks through 4 layers of cheesecloth which have been previously fixed in the necks of 2-liter Erlenmeyer flasks and sterilized in situ . After filtration, the filter is replaced by a sterile cotton plug and the flask is placed on the shaker. Ten filtrations are carried out, at 5, 8, 10, 12, 15, 18, 21, 24, 31, and 42 hours after inoculation, respectively. Platings are made from thc 8th, 9th and 10th filtrations.

(6) Following the 8th and 9th filtrations, pipette 25 ml from each culture into a sterile flask and plate 20 aliquots from each sample as described above (if necessary the samples can be stored at 4C and plated the following day). From the last filtration, prepare as many plates as possible. All plates are incubated at 32C for 2 days. (May be stored at 4C until analysis can be carried out.)

(7) Determine the number of colonies and calculate the germination and the survival percentages. The number of viable conidia per ml in the cultures, when started, and following the filtrations, is a rough measure of the effectiveness of the filtration.

(8) Isolate colonies, mainly from the last filtration, in small tubes containing 2 ml of complete medium. For transferring small pieces of colonies, a sharp rigid inoculation needle is used. Incubate the cultures at 32C for about 7 days.

(9) Test the colonies isolated on complete medium for growth-factor requirements by transferring conidia to liquid minimal medium N with a loop-shaped inoculating needle. Keep the two cultures together at 32C for 2 days. Use the isolated colonies which are not able to grow on minimal for further analysis. Calculate the percentage of biochemical mutants.

Preparing cultures: Wild type strain 74A is kept on minimal medium N. Samples from stock cultures are spread on plates of medium N. The hyphae are used to inoculate agar slants. Conidia are spread on minimal crossing medium + sorbose and single colonies are used to inoculate agar slants of medium N. Then, conidia are used to inoculate six 100-ml Erlenmeyer flasks, each containing 20 ml of crossing medium + 2.5% glycerol (no sucrose). Incubate for 7 days at 20C.

B. PRELIMINARY TEST OF BIOCHEMICAL MUTANTS.

To determine the type of growth-factor requirements of the biochemical mutants, preliminary tests are carried out on mixtures of growth factors. In this way mutants can be characterized as: a) amino acid-requiring, b) vitamin-requiring, or c) purine and pyrimidine requiring.

Materials: One petri dish (17 cm in diameter) of each of the following media: synthetic crossing medium + sorbose; the same + sorbose + casamino acids; the same + sorbose plus a vitamin mixture; the same + sorbose + hydrolyzed yeast nucleic acids. 10 small tubes with complete medium N. 0.9% NaCl and centrifuge tubes. 5 cultures of the biochemical mutants isolated in the filtration experiment. 5 cultures of biochemical mutants with known requirements.

Experimental Procedure: Prepare a suspension of conidia from each mutant by adding 2 ml of 0.9% NaCl to each tube and stirring briefly. Pipette the suspension into a centrifuge tube and adjust the volume to 10 ml with 0.9% NaCl. Centrifuge at 3000 x g for 3-5 minutes. Decant and resuspend conidia in 10 ml 0.9% NaCl. Repeat 2 times. The final volume should be about 5 ml. Prepare a new culture from each mutant on complete medium N. Using an inoculating loop, place a small drop of the conidial suspension of each mutant on each of the media listed above under materials. Before inoculation, mark all plates in such a way that 8 of the mutants are placed at the periphery and 2 in the center. Incubate the plates at 25C for 2 days (can be stored at 4C). Classify the mutants on the basis of their growth factor requirements.

C. AUXANOGRAPHIC TEST:

Prepare a suspension of conidia from each of the mutants and inoculate, as described above, plates with individual growth factors. Test the amino acid-requiring mutants on plates of minimal crossing medium + sorbose and supplemented, singly, with 150 mg/ml of the following amino acids: L-arginine HCl,

L-leucine, lysine, methionine, phenylalanine and proline. Test the mutants requiring purines or pyrimidines on minimal crossing medium + sorbose +adenine sulfate 150 mg/ml. Test the vitamin-requiring mutants on minimal crossing medium + sorbose supplemented with pyridoxine 50 mg/ml and with Ca-pantothenate 50 mg/ml, separately. On a normal sized petri dish, 4 mutants can be tested. Inoculate close to the periphery of the dishes at previously marked positions. Incubate at 25C for 2 days. Many of the auxotrophic mutants isolated by the filtration technique described can be classified by this auxanographic test.

References: Emerson 1955 Vol. 2, pt . 2, p. 443 In Hoppe-Seyler and Thierfelder (ed.) Handbuch der physiologisch- und pathologisch-chemisches Analyse, 10th ed. Springer, Heidelberg; Horowitz 1950 Adv. Genet. 3:33; Pontecorvo 1949 J. gen. Microbiol. 3:122.

STRAINS USED AND PREPARATION OF CULTURES:

The following strains with known requirements are used: pro-1 (21863), FGSC#29; phen-1 (H3791), FGSC#504; me-8 (P53), FGSC#98; me-5 (9666), FGSC#140; arg-1 (B369), FGSC#324. Cultures are kept on complete medium. Stock cultures are maintained and tested as described for arginine mutants.

D. PRECURSOR TEST OF ARGININE MUTANTS.

The final step in the physiological characterization of the biochemical mutants is the precursor test. A group of mutants with known requirements for arginine is used to demonstrate the intermediate steps in the biosynthesis of arginine in Neurospora.

Materials: Cultures of 7 arginine mutants. Plates of minimal medium supplemented as described below. 0.9% NaCl. Centrifuge tubes.

Experimental Procedure: Prepare a suspension of conidia from each of the mutants by adding 2 ml of 0.9% NaCl to each culture tube. Stir briefly and pipette about 1 ml of the conidial suspension into a centrifuge tube. Adjust the volume to 10 ml with 0.9% NaCl. Centrifuge at 300 x 9 for 3-5 minutes. Decant and resuspend conidia in 10 ml 0.9% NaCl. Repeat 2 times. The final volume should be about 5 ml. With a loop a small drop of conidial suspension is placed on the following media:

minimal crossing medium + sorbose; the same + L-arginine 150 mg/l; the same + L- citrulline 150 mg/l; and the same + L-ornithine 150 mg/l. Mark the plates and inoculate at the indicated spots at the periphery. Incubate at 25C for 2 days.

Determine the block in the biosynthetic chain for each of the mutants from their growth patterns. Establish the sequence of the arginine biosynthetic pathway.

Cultures and Maintenance: The following strains of arginine mutants should be used: arg-1 (36703T), FGSC#273; arg-1 (B369), FGSC#324; arg-1 (46004), FGSC#528; arg-5 (27947), FGSC#274; arg-5 (27974), FGSC#480; arg-3 (30300), FGSC#1068; and arg-3 (30300), FGSC#1069. The strains are kept on Vogel's minimal medium N +arginine 150 mg/l. Stock cultures are covered with sterile paraffin oil as soon as conidia are formed and are placed at 4C. Samples from a stock culture are taken with a loop and spread on plates. Hyphae formed on the plates are used to inoculate agar slants. Conidia should be tested on liquid minimal medium N for back mutations. If back mutation has occurred, conidia should be spread on minimal medium N + sorbose + arginine and single colonies re-isolated onto agar slants. Each culture must then be tested on liquid minimal medium. Incubate cultures at 25-32C.

Media:

1) Synthetic crossing medium (Westergaard and Mitchell 1947 Am. J. Botany 34:573). A 4x concentrated stock solution is prepared. After dilution sugar is added and the pH adjusted with NaOH.
2) Vogel's minimal medium N (Vogel 1956 Microbial Genet. Bull. 13:42). A 50x concentrated stock solution is prepared. After dilution sugar is added and the pH is adjusted with NaOH.
3) Complete medium N. To each liter of medium N add: yeast extract 2.5 g., malt extract 5 g., vitamin mixture 10 ml, hydrolyzed nucleic acid 2 ml, and casamino acids 15 ml.

4) Vitamin mixture. Dissolve the following in 1 liter of water: thiamine 100 mg, riboflavin 50 mg, pyridoxine 50 mg, Ca-pantothenate 220 mg, p-aminobenzoic acid 200 mg, nicotinamide 200 mg, choline HCl 129 mg, inositol 400 mg.
5) Hydrolyzed nucleic acid. Dissolve 1% yeast nucleic acid and 0.2% thymus nucleic acid in 1 N NaOH. Hydrolyze for 24 hours at 35C. Adjust to pH 6.5 and store at 4C.
6) "Synthetic casein" (casamino acids). In 687 ml of water, dissolve the following: glycine 20 mg, DL-alanine 180 mg, L-leucine 243 mg, DL-isoleucine 485 mg, DL-aspartic acid 410 mg, L-glutamic acid 1090 mg, DL-serine 580 mg, DL-threonine 390 mg, L-proline 400 mg, L-hydroxyproline 10 mg, L-cystine 15 mg, DL-methionine 360 mg, DL-valine 790 mg, DL-phenylalanine 390 mg, L-tryptophan 110 mg, L-tyrosine 325 mg, L-arginine HCl 305 mg, DL-lysine HCl 310 mg and L-histidine HCl 455 mg.
7) Sorbose-containing media. The ordinary carbon source, sucrose, is replaced by a mixture of sorbose, glucose and fructose to give final concentrations of 1.5%. 0.04% and 0.04%, respectively.

Lacy, A. M. Neurospora in the freshman biology course.

The following laboratory procedure and description of necessary materials is taken from a new "Labtext" From one cell to many, by Funk and Lacy 1966 Wm. C. Brown Co., Dubuque, with permission of authors and publisher. Some introductory material has been deleted.

This manual is designed for use in a one-semester elementary biology course devoted to the "cell to organism" approach. Cell structure and metabolism, genetics, plant and animal development, and , finally, the structure of a multicellular animal (the dogfish shark) are dealt with in this course.

The students begin the Neurospora experiment during the second half of the course, after they have each completed a four-week independent project using a microorganism. Consequently, they are reasonably proficient in aseptic technique and use of the microscope. This experiment does not depend for its success on extensive first-hand knowledge of Neurospora on the part of the instructor. Moreover, this experiment, in fact this whole manual, is primarily designed for colleges which wish to present a modern course in biology but which lack both a large staff of graduate teaching assistants and an unlimited supply of specialized equipment. The experiment occupies about one hour of one laboratory period (of which the other two hours are devoted to corn genetics, blood typing and blood cell identification), about 15 minutes of a second laboratory period and about five minutes of a third period.

Materials: (per student) Dissecting microscope with 45-60X magnification

1 glass-handled dissecting needle

1 metal-handled microspatula

1 small flask of 95% alcohol

1 microscope slide

10 (10 x 75 mm) tubes of supplemented agar medium (medium N)

10 (10 x 75 mm) tubes of minimal liquid medium (medium N)

(per 8 students) 2 tubes of a Neurospora cross (biochemical mutant x "morphological mutant")

1 petri dish of 4% water agar

1 wire test tube rack for small (10 or 12 x 75 mm) test tubes

(per class) 1 large water bath, set at 60C

Procedure:

Anatomy of Neurospora: Each group of four students will be given a test tube in which two mutant strains of N. crassa have been crossed. Observe the different structures in the tube with the dissecting microscope. Also, remove some of the organism aseptically, place it on a slide in wet mount, and observe under the compound microscope. Identify the vegetative structures (mycelium, hyphae, asexual conidia) and sexual reproductive structures (perithecia, ascospores).

Analyzing progeny from a Neurospora cross: By isolating random ascospores (haploids) from the cross, allowing them to germinate and grow, and then studying the characteristics of each isolate, it should be possible to determine whether or not the two genes involved are linked together on the same chromosome and, if so, how closely linked they are. For this experiment, each student will isolate and study 10 ascospores. When the ascospores are block and ripe, they are shot out of the perithecia and form small black clumps on the upper side of the test tube. Dip the microspatula into the alcohol, flame, and use it to remove a few spores from the side of the tube. Place them on one end of a 2 cm x 3 cm agar block on a glass slide. Note: agar dries out very rapidly; do not leave this cut piece lying around too long before using. Remember to practice aseptic transfer techniques when dipping into the test tube containing the cross. Aseptic technique is, of course, not possible when working on the agar block.

After flaming your microspatula, cut one side of the agar block into 10 small squares as demonstrated by the instructor. Then, using the glass-handled dissecting needle (DO NOTFLAME), and using the highest power magnification of your dissecting microscope (45-60x), place one black spore on each square of agar. With your flat microspatula, pick up each little agar square individually and place right side up in individual small test tubes of supplemented agar medium. Label each tube with your name, the cross used, and an isolation number.

These small test tubes containing the isolated spores should be placed in a 60C water bath for 40 minutes. This treatment kills contaminants which may have fallen onto the agar during the isolation and all parental mycelia and conidia that may have stuck to the ascospores. It also provides the "heat-shock" necessary for initiation of spore germination. After "heatshocking", the tubes should be incubated at room temperature until the next laboratory period (or for at least 4 days).

During the next laboratory period, the following procedures should be carried out: 1) The germination percentage should be determined. 2) The tubes containing germinated spores should be examined and the morphology of the organism in each tube recorded. The instructor will demonstrate to you the different types of morphology you may expect to observe. If one parent in the cross was an albino mutant, set the tubes in the light for at least one hour before classifying. Carotenoid synthesis in Neurospora is light dependent. 3) A few conidia or small bits of the mycelium from each tube should be aseptically transferred to tubes of minimal liquid medium (a medium which will support the growth of wild type Neurospora, but not of biochemical mutants).

These inoculated tubes of minimal medium should be incubated at room temperature for two or three days. Then the following points should be determined. 1) Which tubes showed growth and which did not? 2) What does this tell you about the phenotypes of these progeny? Genotypes? 3) What is the correlation between the morphological characteristics of the progeny and the biochemical characteristics (as determined by growth tests on minimal medium)? 4) Is the gene locus determining the morphological character linked to the one determining the ability to synthesize a certain nutrient factor? At the end of the laboratory period, pool your data with that of others in the class studying the same cross, so that you can determine with some statistical accuracy whether or not the two genes are linked.

Before the ascospores are shot out of perithecia, they are contained in sacs called asci (eight spores to each) and are arranged in order of meiotic segregation within these asci . Why do people often isolate spores in order from the asci? If you wish, you may try to do this yourself. Remove a ripe, black perithecium from the crossing tube, squeeze it open with forceps, pull an ascus with eight ripe, black spores to a clean part of the agar, and dissect in order with a microneedle. This requires a steady hand.

Preparation of Materials: The glass-handled dissecting needles are made by inserting 00 Genuine Bohemian Insect Pins (Carolina Biological Supply Co. #A740) into the melted ends of hollow soft glass rods (diam. 5 mm; length about 6 inches) and then flattening the glass onto the pin with forceps. The needles are most easily stored by passing them through a rubber stopper and plugging the stopper into an 18 x 75 mm test tube.

The metal-handled microspatula is made by hammering a 3.5 inch piece of nichrome or chromel wire (24 gauge) until the tip is more or less square, cutting the edges off sharply, and inserting the wire into any fairly short inoculating loop holder.

Test tube racks can be made by buying wire fencing of the appropriate mesh, cutting into two rectangles and a square, and weaving or welding the broken ends together to make a three-tiered rack.

Minimal media for Neurospora are available from Difco (0324-15, 0460-15, 0817-01). We have not tried these, but they would probably be convenient for course work. The most commonly used vegetative medium for Neurospora is medium N or Vogel's medium (Vogel 1956 Microbial Genet. Bull. 13: 42). This is generally made up 50x and diluted as needed.

The supplements added to the medium will depend upon the mutants used. 150 ug/ml of L-amino acid is usually sufficient for growth of any amino acid mutant, but for crossing and germination higher concentrations in the range of 200-300 ug/ml are often more satisfactory.

Neurospora strains of all kinds are available free of charge (in limited numbers) from the Fungal Genetics Stock Center, Dartmouth College, Hanover, New Hampshire. Albino is one of the most satisfactory morphological mutants for class use. Almost any amino acid mutant is suitable for the biochemical mutant. We usually use albino and a tryp-3 mutant as parents. Be sure to order parents of different mating types (A and a).

Neurospora requires a special medium for crossing. A synthetic crossing medium (Westergaard and Mitchell 1947 Am. J. Botany 34: 573) is generally used in research, but corn meal agar with dextrose (Difco B114) is probably simpler to use for course work. The best procedure is to inoculate one parent onto a slant of corn meal agar and incubate at 25C for five days. At that time, dust the conidia from the second parent over the mycelium of the first and reincubate at 25C. While two weeks should be sufficient to obtain ripe spores, it is wise to start crosses a month or two before needed. When ripe black ascospores appear in clumps on the upper inside surface of the tube, the cross may be stored in the refrigerator until needed. Refrigerated crosses will retain high viability for as long as a year.

Lacy, A. M. Neurospora in the genetics course.

While other organisms are used (with some reluctance) in my genetics course, much of the laboratory time is devoted to a long project using Neurospora crassa. The following procedure, which will probably be published in more detail as part of a projected Genetics Laboratory Manual, is designed for use in a course taken primarily by junior biology majors and pre-med students. All of these students have completed the elementary biology course, in which Neurospora is introduced, and most of them have also completed a year of organic chemistry before taking the genetics course.

This type of project is probably feasible only where the number of students enrolled in the lab is less than 30, and where Neurospora research is routinely done in the department. While we have limited our study to tryptophan mutants, this is, in part, a reflection of our own research interests. Presumably mutants involving any well-investigated pathway could be used. About 8 weeks of a 10-week term are necessary for completion of the project. By no means all of the laboratory time is devoted to Neurospora, but 8 weeks' total time is necessary to allow growing time for several repeats of various tests, especially the intra-genic complementation tests.

The general approach involves:

1) Induction (by UV or nitrous acid) and selection (by filtration and selective plating) of tryp mutants.
2) Determination of the particular tryp locus altered in each mutant.
3) Assignment of 3 or 4 tryp mutants (at least 2 of these tryp-3 mutants) to each student. If the yield of new tryp-3 mutants is small, the supply is augmented by "unknowns" from our stock collection.
4) Characterization of the tryp mutants in relation to each other and to known tryp-3mutants on which published data are available. This procedure includes tests for

temperature sensitivity, indole and/or indole-glycerol accumulation, ability to grow on indole, linkage to fluffy, and intra-genic complementation.

5) Assay of crude extracts of a wild type and of a tryp-3 mutant for reaction 2 activity of tryptophan synthetase. For this purpose one batch of wild type strain and one of a tryp-3 strain are grown and the lyophilized mycelium distributed to small groups of students working together on the assay.

6) Preparation of a report by each student. This report is written in the form of a scientific paper and includes the characteristics of the mutants studied, how the mutants are related to previously known mutants, possible explanations for conflicting results, etc.

By setting up this project in such a manner that students are studying unknown mutants within a partially structured procedure, it is possible to allow for individuality of evaluation and speculation and for some freedom of experimentation without requiring the additional teaching staff and the variety of specialized equipment that would be necessary to allow a completely free project.

To present the procedure in detail would be to publish a whole lab manual in the NN. Moreover, the exact procedure differs somewhat from year-to-year, partly in response to availability of equipment, strains, etc., and partly in response to the instructor's need for variety. If any reader desires an amplification of some aspects of this project, I can sometimes be reached by mail, but more surely by telephone. As an example, the procedure by which the students obtain mutants for use in this course is given below.

NEUROSPORA MUTANT HUNT

When Neurospora (or any other organism) is exposed to a mutagenic agent (such as ultraviolet light, x-rays, nitrous acid, etc.), mutations may occur in many different genes in many different cells. If we wish to keep only the mutated types, especially if we wish to collect only certain of the mutated types, an effective selection technique is required. In this course we will use a local modification of the Woodward and Srb filtration method (1954 Proc. Natl. Acad. Sci. U. S. 40: 192). This method is based on the fact that wild type Neurospora can grow in a simple medium containing only minerals, biotin, and sucrose, while nutritional mutants cannot. After treatment with a mutagen, the conidial suspension is incubated in liquid minimal medium. The wild type conidia will germinate, grow, and can be filtered off; the ungerminated mutant conidia will pass through the filter. After repeated cycles of incubation and filtering, the relative proportion of mutant conidia remaining in the medium will increase.

The filtrate is then distributed in petri dishes containing minimal medium supplemented with sorbose (which causes Neurospora to grow in a compact, pellet-like form) and with nutrients required by the desired mutant type. In an ideal experiment, a large proportion of the little colonies which appear on the plates will be of the desired type.

The following procedure will be used for induction and selection of tryptophan-requiring mutants from wild type 74A.

1) Each group of students will be given 5 slants of wild type strain 74A (grown for 5 days on minimal agar medium). Vogel's minimal medium N is used throughout these experiments.
2) Pour 5 ml of sterile distilled water into each of the 5 slants. Disperse the conidia, using a sterile microspatula.
3) Pour suspensions aseptically through a sterile glass wool filter apparatus until about 15 ml of suspension has collected under the filter. A calcium chloride drying tube (the top plugged, the stem wrapped in cotton and inserted in a test tube, and the bulb lined with glass wool) is used for this purpose. Remove the filter and stopper the tube of filtrate with a sterile cotton plug.
4) Pipette, aseptically, one drop of well-shaken suspension onto each side of a haemocytometer. Count the number of conidia in each of 5 "big-squares-containing- 16-little-squares" (demonstration). Average the 5 tallies and calculate the number of conidia/ml of suspension. The most efficient concentration of conidia will differ somewhat with the mutagen to be used. For UV treatment, 10^7-10^8 conidia/ml is desirable.

5) The induction of mutations:

   A. Chemicals. The appropriate concentration of mutagenic chemical is added directly to the conidial suspension and incubated for a given length of time to obtain the desired ratio of mutated to killed nuclei. The action of the chemical is then stopped, usually by the addition of a second chemical which neutralizes its action or by dilution of the mutagenic chemical.

   B. Ultra-violet light. Pipette, aseptically, 11- 12 ml of conidial suspension into an empty sterile "deep-dish" petri dish. Quickly replace glass cover. Warm up UV lamp for about 10 minutes. (Do not look at the lamp bulb - the rays can be very damaging to your eyes.) Place the petri dish under the lamp 10 cm below the bulb. (It is desirable to wear rubber gloves while working under the lamp.) The UV source in this experiment is an 8 watt germicidal UV lamp. When you are ready to irradiate your suspension, remove the top of the petri dish (UV rays will not penetrate glass) and, holding the bottom of the dish between thumb and forefinger, rotate gently to obtain maximum exposure of conidia. Time the irradiation and replace the glass petri dish cover (which has been held face down to maintain asepsis). The class will be divided into 5 groups: members of these groups will irradiate their suspensions for 0, 1, 1.25, 1.5, 2 and 3 minutes, respectively.

6) Pipette 5 ml of treated suspension into each of 2 sterile 1-liter Erlenmeyer flasks with filter tops containing 250 ml of minimal medium. Add 0.25 g of streptomycin sulfate " as aseptically as possible". The filter-top plugs are made by placing a rectangular strip of curity 60 cheesecloth across the mouth of an Erlenmeyer flask, then placing the same sized strip across the mouth at a right angle to the first, and then plugging the flask with cotton so that the cheesecloth is pushed down into the flask but the ends are left above it. This makes it

possible to remove the plug easily and yet be left with a layer of cheesecloth for filtering.

7) Label flasks with all pertinent information and incubate flask X at 25C and flask Y at 30C.

8) Now for the filtration part of the procedure. The most efficient way of handling the flasks for rapid and aseptic transfer from a flask of liquid, through a sterile cheesecloth filter, and into a sterile empty flask will be demonstrated. The 0 time suspension should be filtered at approximately 9 hours after inoculation. Both the 0 time suspension and the treated suspensions should be filtered at approximately 8 AM, 2 PM and 9 PM for the next two days. Probably one filtration each is sufficient for the 3rd and 4th days. Note: 24 hours after inoculation, filter into flasks containing 100 ml of sterile minimal medium. All other filtrations should be into empty flasks.

9) The time of filtrate plating (usually after 5 days of filtration or when 12 hours passes with no formation of new mycelia) will be announced in class. Small aliquots of filtrate (the exact amount depending on the concentration of treated suspension and the length of time treated) will be pipetted, aseptically, into flasks of molten, but fairly cool (approximately 45C) minimal sorbose agar (1% sorbose, 0. 1% sucrose) containing (since we are selecting for tryptophan requiring mutants) about 25ug L-tryptophan/ml. The agar suspension will be distributed over 10 already-layered plates for each X and each Y flask, and the plates will be incubated at 25C and 30C, respectively.

10) The plates should be inspected at least once daily (preferably twice) and any visible spots of growth cut out of the agar (aseptically) with a microspatula, inoculated into small tubes of tryptophan-supplemented agar, and incubated at 30C until good growth is obtained (about 3 days). Draw a crayon circle on the petri dish bottom around the spot picked so that you and your lob partners do not repick the some colony later.

11) Transfer, aseptically, tiny wisps of conidia from each agar slant to appropriately labelled small tubes of

minimal liquid medium; make two tubes per presumptive mutant. Incubate one at 25C and the other at 37C. Save presumptive mutant stocks in the refrigerator.

12) Those isolates which DO NOT show significant growth on minimal medium at bothtemperatures after 2 days' incubation can be assumed to be mutated in some gene controlling tryptophan formation. Why? Check with the instructor as to what constitutes "significant growth". The mutant stocks should be numbered, transferred onto 3 large tryptophan-supplemented slants, and incubated at 30C.

13) At this point, the tryptophan mutants obtained by the class will be supplemented by tryptophan mutants obtained by the instructor, so that each student will have 3 unknown tryptophan mutants. Study and characterization of these mutant strains will constitute the central experiment of the term's laboratory work.

The exercise given below is used in the laboratory portion of a general genetics course. Students work singly or in groups, depending upon the facilities available. If desired, the students can be assigned the duty of preparing reagents for this exercise during the preceding laboratory period.

Materials Needed:

Equipment: Table top centrifuges (capacity of at least 1500 rpm) and centrifuge tubes incubator or waterbath adjustable to both 37 and 60C

125 ml Erlenmeyer flasks

test tubes and test tube racks

graduated cylinders

glass stirring rods

glass-stoppered bottles

Reagents:

1) 0.1 SCC - A stock solution of 10x SCC (sodium saline citrate) can be prepared by mixing 87.5 g NaCl and 44.1 g Na citrate and diluting to 1 liter.

2) Saline EDTA - Add 8.75 g NaCl to 37.23 g disodium EDTA and dilute to 1 liter. Adjust pH to 8.0 with NaOH.

3) Sodium lauryl sulfate - Dissolve 25 g in 75 ml distilled water. Note: solution solidifies below 25C.
4) Tris buffer - Mix 32.5 ml of 0.1 N HCl with 25 ml of 0.2M solution of tris (hydroxymethyl) amino methane and dilute to final volume of 100 ml.
5) Acetate EDTA - Mix 40.8 g Na acetate with 0.037 g EDTA and dilute to 100 ml.
6) RNAse - Obtain commercial RNAse (Calbiochem) and place water solution of this enzyme (2 mg/ml) into boiling water for 10 minutes to destroy DNAse activity.
7) Phenol reagent - Saturate phenol with saline EDTA and adjust pH to 8.0 with 10N NaOH. Caution should be exercised in the preparation and use of this reagent.
8) Isopropanol.

Neurospora powder: Neurospora powder can be produced by harvesting mycelia grown in liquid medium (for a period of 14-18 hours) and then lyophilizing in a freeze dryer. The dry mycelia may be powdered by passage through a Wiley mill (mesh size 60). If facilities are unavailable for the production of Neurospora powder, it can be purchased. The powder may be stored in the freezer for long periods of time.

Isolation Procedure:

1) Mix 1 g lyophilized Neurospora powder with 25 ml saline EDTA and 2 ml sodium lauryl sulfate in a 125 ml Erlenmeyer flask.
2) Stir by hand or by use of magnetic stirrer at low speed.
3) Incubate in 60C incubator for 30 minutes.
4) Remove from incubator and add 1-2 drops of chloroform.
5) Transfer to a stoppered bottle and add an equal volume of phenol reagent. Slowly rotate for 3-5 minutes.
6) Pour into centrifuge tubes and centrifuge at 1200-1500 rpm for 10 minutes.
7) Remove supernatant with pipette attached to aspirating tube.
8) Carefully layer two volumes of ethanol on the surface of the supernatant.

9) Carefully spool out DNA with a glass rod and redissolve in small quantity of 0.1 SCC.
10) Adjust pH to 7.8 with Tris buffer.
11) Add RNAse to a concentration of 50 ug/ml DNA solution and incubate for 30-60 minutes at 37C.
12) Remove from incubator and add 1 ml acetate EDTA.
13) Carefully layer approximately 7 ml isopropanol on the surface and spool out the DNA.
14) Redissolve DNA in 0.1 SCC and purify by deproteinization with phenol as previously described.

Using these procedures, DNA of high molecular weight and good purity can be obtained. If desired, the identification and quantitation of the isolated substance can be made by means of UV-spectrophotometry or chemical tests, such as the diphenylamine reaction. For a 2-hour laboratory period, the procedure can be carried through step 8. The product at this stage is DNA with adhering protein and RNA. The experiment may be terminated at this point or else the DNA can be dissolved in 0.1 SCC and stored in the refrigerator and the isolation can be continued during the next laboratory period. Haskins, F. A. Accumulation of anthranilic acid by a mutant strain of Neurospora. The laboratory exercise given below has been used successfully for a number of years in a genetics course taken primarily by graduate students, but could readily be used in an undergraduate course, if desired.

Preliminary Statement: Some Neurospora mutants accumulate metabolic intermediates in the mycelium or in the surrounding medium. The indentification of such intermediates provides important clues concerning the nature and the genetic control of certain biosynthetic pathways. The object of this exercise is the demonstration of the accumulation of anthranilic acid by a tryptophanless mutant of Neurospora. Anthranilic acid is an intermediate in the biosynthesis of tryptophan (Wagner and Mitchell 1955 Genetics and metabolism, p. 310. 2nd ed. Wiley, New York).

Strains Used: Strain 10575. This strain is defective at the tryp-1 locus. It grows if supplied indole or tryptophan, but is not able to utilize anthranilic acid for growth.

Strain B1312. Judging by its growth responses, this strain appears to be mutant at the tryp-2 locus. Other tryp-2 mutants (e.g. 75001 or 40008) would probably be equally useful in this experiment. Mutants of the tryp-2 type are able to utilize anthranilic acid, indole, tryptophan, or certain other related compounds for growth.

Both mutant strains are readily cultured on slants of Fries' minimal agar supplemented with L-tryptophan at a concentration of 0.1 mg/ml. Cultures approximately 1 week old should be available for the required inoculations.

Equipment and Supplies:

Fries minimal medium. (Other minimal media commonly used for Neurospora would probably be equally satisfactory).

Anthranilic acid and L-tryptophan. At a concentration of 1 mg/ml these compounds dissolve quite readily in water. Warming speeds solution. Dilute anthranilic acid to 0.2 mg/ml for convenience in bioassay.

Sterile water. Add 1 or 2 ml of water to each of several 4-inch test tubes, plug with cotton, and autoclave.

Pasteur pipettes. These are conveniently made by cutting 7-mm soft glass tubing into 15 cm lengths, plugging both ends of each length with cotton, and autoclaving. Shortly before use, heat the center of each length to softness in a Bunsen flame and pull out to form 2 pipettes.

Erlenmeyer flasks, 125-ml. Nine flasks will be needed for each student or pair of students.

Pipettes, 1-ml, graduated. Each student, or pair, needs 2.

Graduated cylinders, 25-ml . Each student, or pair, needs 1

Chromatography paper, 6 x 11-inch sheets of Whatman No. 1. Several students may use the same sheet of paper.

Chromatography vessels. One is needed for each sheet. Gallon jars may be used, but standard 6" x 12" chromatography jars are more convenient. Saranwrap is satisfactory for covering the jars.

Solvents for chromatography, about 100 ml per vessel. Two freshly prepared solvents are needed:

n-propyl alcohol - 1% ammonia (2:1, v/v) and

n-propyl alcohol - 1% acetic acid (2:1, v/v)

Glass capillaries. These may be drawn out from 7-mm soft glass tubing.

Ultraviolet lamp or "Blacklight". Peak emission of the lamp should be near 360mu. Autoclave.

Drying oven, 80C.

Analytical balance.

Procedure: Prepare, and plug with cotton, the following flasks; a) 20 ml minimal medium, b) 20 ml minimal + 0.25 mg anthranilic acid, and c) 20 ml minimal + 0.25 mg L-tryptophan. Autoclave the flasks 20 min. at 15 psi, cool, and inoculate each flask with 1 drop of a suspension made by dispersing a small amount of 10575 conidia in sterile water. Incubate the flasks at room temperature or in a 25C incubator if one is available. The response of strain 10575 to minimal medium, to anthranilic acid, and to tryptophan is shown by the growth or lack of growth occurring in the 3 flasks. The medium in flask c, after growth of strain 10575, will be used in next week's laboratory.

After an incubation period of 7 days (4 to 7 days should be satisfactory, depending on the laboratory schedule), observe the 3 flasks under the ultraviolet lamp. The blue fluorescence of the medium in flask c is produced by a number of substances, perhaps the chief of which is anthranilic acid. Proof of the identity of this compound would require its isolation in pure form from the medium. Such isolation will not be attempted as a part of this exercise, but a partial characterization and assay of the compound will be made on the basis of paper chromatography and biological activity.

A. Paper chromatography. Using a glass capillary, apply a spot of the medium from flask c to a previously marked position on a line 3/4 inch from one of the long edges of a 6 x I 1-inch sheet of Whatman No. 1 filter paper. Sufficient medium should be applied to make the spot approximately 1/4 inch in diameter. Let the spot dry, then repeat the application. Continue this sequence of spotting and drying until 10 applications have been made at the same position. As a control, make one application of the anthranilic acid solution having a concentration of 1 mg/ml . The control spot should be applied along the base line about 1 inch from where the medium was applied. Control spots of additional fluorescent compounds may be used if desired. (Also, the chromatographic separation of the constituents of washable black Skrip ink makes a rather striking demonstration which can be observed while the solvent is traveling up the paper.) After all spots are dry, staple together the ends of the sheet to form a cylinder 6 inches high. Do not permit the ends to overlap each other. Place the cylinder in one of the solvent vessels provided, and cover the vessel. Part of the class will use the n-propyl alcohol-ammonia solvent and the remainder will use the n-propyl alcohol-acetic acid solvent. After approximately 2 ½ hours, remove the chromatograms from the solvents, dry, and examine them under the ultraviolet lamp. Mark the solvent front and any fluorescent spots that are apparent. Measure distances from the base line to the solvent front and to the centers of any fluorescent spots. Calculate the Rf value of each spot as the ratio of the distance traveled by the solute to the distance traveled by the solvent. Each student should observe and record results obtained with both solvents.

B. Bioassay. Prepare and autoclave the following flasks: 1) 20 ml minimal medium, 2) same, + 1 ml medium from flask c, 3) same, + 0.02 mg anthranilic acid, 4) same, + 0.05 mg anthranilic acid, 5) same, + 0. 10 mg anthranilic acid, and 6) same, + 0.20 mg anthranilic acid. Inoculate each of the 6 flasks with 1 drop of a conidial suspension of strain B1312 (or other tryp-2 mutant). Incubate at room temperature (or 25C) for 4 days, then harvest the mycelial pads, squeeze out most of the liquid, dry the pads overnight at 80C, and weigh them to the nearest mg. Plot dry weights against quantity of anthranilic acid, for flasks 3, 4, 5, and 6. Assuming that anthranilic acid is the only material in 10575 culture filtrates with activity for strain B 1312, calculate the quantity of the compound present in flask c. (Students typically obtain values approximating 0.15 mg of anthranilic acid/ml, or a total of about 3 mg in flask c.

THE ROLE OF OROTIC ACID IN PYRIMIDINE BIOSYNTHESIS IN NEUROSPORA.

The experiment below has been used in an advanced biochemistry course for the past four years with considerable success. Part A normally takes three full laboratory periods, starting on a Friday to allow for growth of cultures over the weekend. Part B was adapted from a similar experiment with E. coli described in Cowgill and Pardee (1957 Experiments in biochemical research technique. Wiley, New York)

Objective: The experiment provides experience in radioactive tracer methodology, in the use of mutant organisms, and in the isolation and characterization of compounds of biological significance.

Equipment and Supplies:

Cultures: N. crassa wild type strain 1A and mutant strain pyr-4 (36601).

Medium: Fries basal medium.

Isotope: 6-^{14}C-orotic acid (New England Nuclear Corporation).

Equipment: Flash evaporator

UV spectrophotometer

Clinical centrifuge

Hood

Autoclave

Incubator-shaker

Planchet-type radiation counter

Fraction collector (optional)

Mineralite UV lamp

Chromatography tank

Procedure:

Part A. Incroporation of 6-^{14}C-orotic acid into ribonucleic acids in Neurospora.

Growth procedure: The mold is grown in 125 ml Erlenmeyer flasks. 1 ml of a stock solution of orotic acid-6-^{14}C containing 2.5 x 10^6 cpm/ml is added directly to 125 ml of Fries medium by means of a pipettor. Unlabelled orotic acid (40 mg) is added separately to the medium, and 12.5 ml volumes are pipetted (plugged pipettes) into each of ten flasks. The flasks are then stoppered with cotton plugs and autoclaved for 20 minutes. After cooling, the flasks are inoculated with 0.2 ml of a conidial suspension of the mold, made by dispersing two loopfuls of the conidia in 10 ml of sterile distilled water. Incubation is at 25C for 3 days, by which time conidiation is just commencing. The contents of the flasks are then filtered on a fritted glass funnel with suction and the mycelial residue is washed with 5 ml of water and then soaked in 100 ml of acetone for 15 minutes. The acetone is removed and the mycelium washed with dry ether and allowed to dry. A brittle disc is thus obtained which is suitable for grinding.

Isolation of ribonucleotides: The mycelium obtained from 10 flasks of the wild strain of the mold grown for 3 days on orotic acid-6-^{14}C is ground in a mortar with 120 mesh carborundum powder for 15 minutes. The resulting powder is extracted 3 times with 10 ml of cold, 10% (w/v) trichloroacetic acid, transferring to a 12-ml Pyrex centrifuge tube in the process. The solid residue is washed once with 4:1 (v/v) ethanol-water and then extracted 3 times with boiling 3:1 (v/v) ethanol-ether. The extractions are carried out by suspending the centrifuge tube in a boiling water bath and stirring the contents occasionally. The process is carried out for a half-hour the first time and for 6 minutes the remaining times. The lipid-extracted residue is then washed twice with ether and air dried. Five ml of 1N KOH is then added, and the mixture allowed to stand at room temperature for 24 hours.

After centrifugation, the supernatant is transferred to a second 12-ml centrifuge tube, cooled on ice, and acidified to a pH of 3 with concentrated perchloric acid (pH test paper). The resulting precipitate of potassium perchlorate and protein is allowed to coagulate for 10 minutes and then removed by centrifugation. The resulting supernatant of ribonucleotides and suspended protein is filtered through a layer of Celite on a fritted glass filter and brought to a pH of 11 with 1N KOH. The solution of ribonucleotides is thus obtained and is allowed to filter into a 1 x 27 cm column containing Dowex 1 anion-exchange resin (Cl- form) of 200-400 mesh size and 10% cross-linking. The column is developed first with 200 ml of water and then with 200 ml of 2N HCl. The optical density of the latter eluate is determined at 260 mu, and an approximate molar extinction coefficient of 10,000, together with an average molecular weight of 350, is used to estimate the concentration of mixed nucleotides (ca. 5 mg). The solution is then evaporated in vacuo to dryness several times to remove hydrochloric acid, yielding a greyish-white residue.

Hydrolysis of ribonucleotides: The mixed ribonucleotides are taken up in several ml of 0.1N HCl and transferred to a glass-stoppered, 10-ml volumetric flask. The mixture is then blown to dryness by means of a stream of charcoal-filtered air. After careful addition of 0.5 ml of concentrated perchloric acid, the flask is heated on the steam bath behind an explosion shield for 40 minutes. The contents of the flask are then transferred to a 12-ml centrifuge tube with the aid of several small portions of water, and, after centrifugation and transfer of the supernatant to a second graduated centrifuge tube, the precipitate is washed and the washings added to the tube. The solution is then diluted to 5 ml, to provide a perchloric acid concentration of about 1N, preparatory to separation of the mixed bases.

Isolation of the purine and pyrimidine bases: The solution of bases in 1N perchloric acid is allowed to filter into the resin bed of a 1 x 27 cm Dowex 50 column (hydrogen form) 200-400 mesh. Elution with 100 ml of water, collecting 10-mi fractions, serves to remove uracil mixed with perchloric acid. Cytosine usually comes off in an 80-ml volume after ca. 160 ml of 2N HCl has been passed through the column. Guanine is next eluted by 3N HCl in an approximately 120-ml volume after about 60 ml of the eluant has passed through the column, and adenine is then removed by ca. 140 ml of 4N HCl after a forerun with about 60 ml of the acid. Elution should be followed by means of the Beckman spectrophotometer and the OD at 260 mu and at max plotted against the volume collected.

The uracil-perchloric acid solution is carefully taken to 3 ml volume and adjusted to a pH of 11 with 5N KOH. After removal of the potassium perchlorate precipitate, the solution is placed on a 1 x 10 cm Dowex 1 column in chloride form. After the solution has filtered into the resin bed, 100 ml of 0.015N ammonium formate buffer pH 9/1 is passed through, and uracil is then removed in a 75-ml volume after 75 ml of 0.015N ammonium formate buffer of pH 8.0 has filtered through the column. Half-milliliter samples of the separated bases are plated on stainless steel planchets, evaporated dryness and counted. Determine the specific activity of each purine and pyrimidine base. Editor's note: Some of the above procedures involving the use of perchloric acid are dangerous and should be performed by students only under the supervision of an experienced chemist and with the proper safety equipment.

Part B. Isolation of excreted orotic acid from N. crassa strain pyr-4 (36601).

Into a 2-liter Fernbach flask place 700 ml of a medium consisting of Fries basal medium + 0.5 mg/ml yeast extract + 0.1 mg/ml asparagine + 0.05 mg/ml cytidine hemisulfate. Plug the flask and autoclave and, after cooling, inoculate with a conidial suspension of strain 36601. After 3 days' growth, harvest the mycelium, collecting the residual nutrient medium. Store the mycelium at OC. Dilute 0.5 ml of the medium with 2.5 ml of water and read the optical density at 290 mu: the reading should be about 0.2. Assuming that 50% of this absorption is due to orotic acid (e290 = 6.2 x 10^3; mol. wt. of orotic acid monohydrate = 174), how much orotic acid is in the solution? Concentrate the medium to about 200 ml by flash evaporation. Make the solution 0.1N in KOH. After the solution is chilled overnight, potassium orotate should crystallize out. Separate the precipitate by centrifugation, dissolve most of it in as small a volume of boiling water as possible (less than 10 ml), add a little charcoal if much color is still present, and filter while hot. Bring the hot filtrate to 0.1N with HCl and allow orotic acid to crystallize by cooling the solution to OC overnight. Filter, wash the crystals with cold water and dry them in vacuo.

Identify the crystals as orotic acid monohydrate by some of the following criteria: a) melting point 323C (decomposes); b) spectra at several pH (5); c) equivalent weight and pK_a, by titration; d) paper chromatography. For paper chromatography place 5 ul of 0.1% orotic acid neutralized with NaOH on Whatman No. 1 paper, develop the descending chromatogram with a solution containing 50 ml of n-butanol, 15.8 ml of 95%

ethanol, 11.4 ml of formic acid and 22.8 ml of water. Orotic acid gives a spot with an R_f of 0.43 that can be seen under ultraviolet light. Run a sample of authentic orotic acid for comparison. Samples obtained in Part A may also be run against known compounds in this solvent system.

On the basis of the results of Parts A and B, what can you conclude about the role of orotic acid in pyrimidine biosynthesis?

Ishikawa, T. Neurospora as part of an undergraduate genetics course.

The following contribution presents, in outline form, procedures for experiments with Neurospora that are used in a genetics course at the University of Tokyo. The nature of accompanying laboratory lectures is indicated for each experiment, as well. The students, about 15 per class, are undergraduate majors in biology who meet for 5 hours per week for 12 weeks; they are expected to come to the laboratory outside regular class hours to make observations whenever necessary. Media, cultures and sterile glassware for the experiments are prepared by the students, themselves. Several of these exercises may be run simultaneously, depending upon the class schedule.

Experiment 1. Characterization of mutant strains.

Lecture: Mutation, types of mutants, pathway of adenine (AMP) synthesis.

Strains: ad-3; ad-4; ad-6; ad-8; ad-8; lys-5; ylo-1; asco, lys-5; 74A; 3. la.

Procedure:

1) Prepare fresh cultures of mutants and the wild types given. Use glycerol complete slant medium + adenine 100 ug/ml and lysine 100 ug/ml.
2) Sterilize petri dishes, test tubes and pipettes. Prepare sterile water blanks.
3) Prepare media: a) sucrose agar plating medium with and without supplements; b) liquid minimal medium with and without supplements. Supplements: adenine 100 ug/ml, hypoxanthine 100 ug/ml, lysine 100 ug/ml or adenine + lysine.

4) Pour 20 ml agar medium, autoclaved, into each plate. Dispense 1 ml liquid medium into each 10-cm tube, plug autoclave.
5) Observe morphological characters of fresh slant cultures.
6) Make conidial suspensions of each strain in water and shake well.
7) Inoculate conidial suspensions into both media. Mark the strain numbers on plates and tubes and incubate at 25C.
8) Observe growth after 1-3 days. Record morphological and biochemical characters of the strains.

Experiment 2. Allelism test.

Lecture: Discussion of gene, locus, allelism; cis-trans test.

Strains: ad-8 mutants (E6A; E129A; E32A; E111A; E34A; E41A); ad-4A.

Procedure:

1) Prepare fresh cultures of mutant strains given. Use glycerol complete slant medium (+adenine 100 ug/ml).
2) Prepare plates of minimal agar medium and sterile water blanks. Sterilize pipettes.
3) Observe morphological characters of fresh slant cultures. Suspend conidia of each strain in 2 ml water.
4) Mark on the bottom of each minimal agar plate 4 spots on the periphery: one spot for two individual mutants and two spots for the combination of these two mutants. Make all possible mutant combinations.
5) Inoculate the spots marked on the bottoms of the plates with conidial suspensions of the appropriate mutants, using 1-ml pipettes. Incubate at 25C.
6) Observe the results every day for at least 3 days. Consider allelic relationships among the mutant strains given. Construct the complementation map for the allelic mutants.

Experiment 3. Recombination and the genetic map (1).

Lecture: Meiosis, tetrad analysis, recombination, genetic map.

Strains: asco, lys-5; 74A.

Procedure:

1) Prepare fresh cultures of both strains. Use glycerol complete slant medium (+ lysine 100 ug/ml).
2) Prepare slants of crossing medium (+ lysine 100 ug/ml) in 3 x 18-cm tubes.
3) Inoculate conidia of both strains into a slant tube. Incubate at 25C for 2 weeks.
4) Observe at least 100 asci under the microscope and calculate the centromere distance for the asco locus.

Experiment 4. Recombination and the genetic map (2).

Lecture: Linkage, linkage groups, genetic map.

Strains: ad-8, lys-5, ylo-1, a; 74A.

Procedure:

1) Prepare fresh cultures of the mutant and wild type given. Use glycerol complete slant medium (+ adenine 100 ug/ml and lysine 100 ug/ml). Prepare crossing medium, supplemented, in 18-cm tubes.
2) From fresh slant cultures, inoculate mutant and wild conidia onto crossing medium. Incubate at 25C for 3-6 weeks, until ascospores are shot.
3) Prepare plating medium (sorbose minimal agar medium + adenine + lysine), 0.05% agar solution and sterile plates.
4) Prepare ascospore suspension in 0.05% agar solution. Count the number of ascospores in 0.05 ml suspension. Heat shock the ascospores at 60C for 40 minutes in a water bath.
5) Pipette spore suspension (200 spores per plate) into sterile plates. Pour medium (40C) into plates containing ascospores. Mix well. Incubate at 25C for 3 days.
6) Prepare isolation medium (minimal medium + 0.5% agar + adenine + lysine). Melt, pour into 7-cm test tubes, plug and autoclave.
7) Isolate 200 colonies; cut out a piece of colony and transfer it into isolation medium. Incubate at 25C for 5-7 days.

8) Prepare test medium (minimal liquid medium with appropriate supplements poured into 7-cm tubes and sterilized).
9) Observe the characters of each isolate and test for biochemical requirements by inoculating conidia into liquid test medium suitably supplemented.
10) What is the genotype of each isolate? Consider the linkage relationships.

Experiment 5. One gene-one enzyme relationship.

Lecture: Protein synthesis; amylases produced by Neurospora.

Strains: 74A; amylase mutant which shows no amylase activity in the culture filtrate.

Procedure:

1) Prepare fresh cultures of amylase mutant and wild type. Use glycerol complete slant medium.
2) Prepare crossing medium in 18-cm tubes.
3) From fresh slants inoculate mutant and wild type into crossing medium. Incubate at 25C for 4 weeks.
4) Prepare plating medium (sorbose minimal medium).
5) Suspend ascospores in 0.05% agar solution. Count the number of ascospores in 0.05 ml suspension. Heat shock ascospores at 60C for 40 minutes.
6) Plate 200 ascospores per plate. Incubate at 25C for 3 days.
7) Prepare isolation medium (minimal medium in small tubes).
8) Isolate 100 colonies. Incubate at 25C for one week.
9) Prepare: a) minimal liquid medium + 1.5% maltose in 20 18-cm tubes (3 ml/tube), and b) minimal liquid medium + 0.2% starch + 1% sucrose in 100 10-cm tubes (1 ml/tube). Plug and autoclave.
10) Inoculate conidia of each isolate, amylase mutant and wild strain into a and b media. Into a put 20 isolates, and into b put 100 isolates. Procedure b is a convenient simple method to observe the presence of amylase

activity in the culture filtrate. Procedure a is used to confirm the b result. Incubate at 25C; a for 2 weeks, b for 5 days.

11) After 5 days add to the b cultures 0.1 ml of I_2-KI solution. Identify amylase mutants.
12) After 2 weeks, pipette out 0.4 ml culture filtrate from the a cultures and assay for amylase activity. Identify the amylase mutants and compare with the result from procedure b.

Assay Method:

Reaction mixture: to 1.6 ml of 0.11% starch dissolved in 0.05M tris buffer (pH 6.0) add 0.4 ml of culture filtrate. Incubate at 37C for 60 minutes. Stop reaction by adding 0.4 ml 1N HCl.

Starch-iodine reaction: to 1 ml of reaction mixture add 0.5 ml 1N HCl and 1 ml of I_2(0.03%)-KI (0.3%) solution. Dilute to 10 ml by adding 7.5 ml water.

Measure transmittance at 660 mu.

Experiment 6. Induction of mutation.

Lecture: Mutagenesis, forward and back mutations.

Strains: ad-8 (E6A).

Procedure:

1) Prepare a fresh culture of the ad-8 mutant strain. Use glycerol complete medium in 16 100-ml flasks supplemented with adenine, 100 ug/ml.
2) Prepare plating media: a) to detect revertants, minimal sorbose medium, 15 plates for each irradiation, b) for survival test, minimal sorbose medium + adenine 100 ug/ml, 15 plates for each irradiation.
3) Prepare a conidial suspension of the mutant in sterile water. Spin down the conidia in centrifuge tubes. Wash 3 times with water by centrifuging. Resuspend in 90 ml water. Estimated number of conidia should be approximately 1 x 10^8 ml.
4) Pipette 10 ml of conidial suspension into a sterile petri dish and irradiate with a 15W UV-lamp at 30 cm. for

between 0 and 30 minutes, taking samples at 5 minute intervals.

5) Dilute irradiated suspension for survival test, using sterile dilution blanks (9 ml of water) and 1 ml pipettes. Make dilution of 10^{-3}-10^{-6} depending upon the killing rate estimated.

6) Pipette diluted suspensions into plates. Pour in agar medium at 40C and immediately mix well. Plate irradiated diluted suspension onto medium b, using 3 dilutions, 5 plates each. Plate irradiated suspension without dilution onto medium a; 10 plates, 0.8 ml inoculum and 5 plates, 0.1 ml inoculum. Incubate at 25C for 4 days.

7) Count the number of colonies per plate. Draw the survival curve and mutation frequency-UV dose curve.

Experiment 7. Cytoplasmic inheritance.

Lecture: Function of mitochondria, nature of poky mutant, results of reciprocal crosses.

Strains: poky A; poky a; 74A; 3. la.

Procedure:

1) Prepare fresh cultures of mutants and wild type. Use glycerol complete slants.

2) Prepare crossing medium in 18-cm tubes.

3) Inoculate protoperithecial parents: inoculate conidia of all four strains, separately, into tubes of crossing medium. Incubate at 25C for 7 days.

4) Make suspensions of conidial parents in sterile water (all four strains). Pour 0.8 ml of conidial suspension into a tube containing a protoperithecial parent culture. Make all possible reciprocal combinations. Incubate at 25C for 2 weeks.

5) Prepare ascospore suspensions in 0.05% agar solution. Count the number of ascospores in 0.05 ml suspension. Heat shock the ascospores at 60C for 40 minutes.

6) Plate the spore suspensions (500 ascospores per plate) on minimal medium + 1.5% sucrose + 1.5% agar.

7) Observe the size of the colonies after 4 days' incubation at 25C.

Ultraviolet light is one of the most widely studied mutagenic agents. Its mechanism of action has been demonstrated to be a multiple one, consisting of both the formation of photoreversible thymine dimers within the DNA molecule and other, non-reversible alterations of the molecule. This experiment will be carried out both in the dark, thus avoiding photoreactivation, and in the light. When ultraviolet irradiation is carried out in the presence of white light, a considerable portion of the genetic damage induced is repaired through the photoreactivation process.

The second half of the experiment is carried out in almost complete darkness, so it is essential that the students plan and organize the work in advance. Equipment which will be needed during the blocked-out phase should be laid out on the work table in a logical order so that it will not be misplaced.

The student will be furnished a Neurospora culture which has a heavy conidial growth. Suspend the conidia in sterile water by tapping the side of the tube with the index finger. Filter the suspension through a very thin sterile cotton pad into a centrifuge tube to remove the mycelium. Centrifuge the filtered conidial suspension for 10 minutes. Pour off the supernatant and wash the conidia with sterile water, centrifuge and resuspend in sterile water. Determine the concentration of the conidia with a haemocytometer and adjust with sterile water to 5×10^6cells/ml. Irradiate the cells in a petri dish at 50 cm from a 15-watt germicidal lamp for 5 minutes, agitating the suspension constantly while irradiating. Withdraw aliquots at 0, 30, 60, 90, 120 and 300 seconds, diluting each aliquot 10x with Neurospora minimal broth (Difco). Thus, each sample will have been diluted 10 times upon withdrawal from the dish. The platings of each sample, for scoring of survivors, should be as follows: plate 1 ml of the following dilutions, 1×10^{-4}, 5×10^{-3}, 1×10^{-3}, 5×10^{-2}, 1×10^{-2}, 1×10^{-2} and 1×10^{-1}, respectively for the exposure times given above. Pipette the sample into a sterile petri dish and add 10 ml of Neurospora agar (Difco) which has been supplemented with 8 g of sorbose to induce colony formation. The agar should be liquid but not painfully hot to the touch and should be adequately swirled. Incubate the plates right side up at room temperature.

The procedure above should be followed during the second phase of the experiment, except that the lab should be made completely dark to guard against photoreactivation. A yellow safelight bulb may be used, if necessary and if kept at least five feet from the work area.

Construct a graph of survival rate against length of irradiation for Neurospora in both light and dark. Discuss the photoreactivation effect and the effects of ultraviolet light as a mutagenic agent.

The October 1, 1966, letter "To all Neurosporologists", from the editor, was clear in stating the desired emphasis of NN#10, i.e., " the use of Neurospora in teaching", but I am going to take the liberty of expanding that emphasis,. via the catchall "in teaching", to two aspects of laboratory teaching not specified in the letter.

The first point is directed to teachers who may wish to use Neurospora as a laboratory teaching tool but who are at present unfamiliar with it. In my opinion it is unwise to incorporate Neurospora into the student laboratories (a) before the laboratory is properly equipped, and (b) before the teacher has become familiar with, maintained, and worked with the organism. The ill-equipped laboratory coupled with inexperienced hands insures contamination of the cultures and a teacher unfamiliar with the eccentricities of Neurospora may only add to the confusion of interpreting experimental results. If Neurospora is used under these conditions, it would be best to "mark" all the stock cultures (e.g., with albino) so that natural contaminants can be screened . My experience leads me to conclude, however, that Neurospora is not an ideal organism for the amateur who has neither the time, equipment nor inclination to become thoroughly familiar with it, and that the confusion created by contamination and unfamiliarity in beginning laboratories for offsets the slim harvest of insight or intellectual stimulation its use may afford.

The second point is directed to teachers who are familiar with Neurospora and who have teaching laboratories equipped to cope with it. These teachers may use Neurospora as a means for designing questions and problems calculated to stimulate the imagination of the student rather than as gimmicks to keep him busy. A plan that I have followed can be adapted to the personality of the teacher and it has proven a worthwhile incentive for my students. In brief, my suggestion is to present to the student the responsibility of becoming acquainted, through the literature, with specific key experiments (each teacher will make his own list), and with some of the questions being asked today by Neurosporologists. In the meantime, the student will do some simple growth kinetics, transferring, ascospore isolations, etc., to become familiar with basic procedures. Following such orientation the student will propose an experimental design, defend the design, modify it to fit all the "feasibles" (time, space, equipment, etc.), and then attempt to execute the experiment.

The isolated pockets of education which permit student participation both in the design and the execution of experiments are rare, and this extends to many graduate schools. It would appear that a well-equipped laboratory coupled with a teacher familiar with the experimental material may well combine to foil the "busy-work" approach taken, usually out of necessity, by teachers with fewer resources, and at the same time take the more positive approach of encouraging student expression.

Bibliography

Adelberg, Edward A.: *Papers on Bacterial Genetics,* Boston, Brown and Company, 1960.

Barnes, N.: *Biology,* New York, Worth Publishers, 1989.

Brandwein, P.F.: *Sourcebook for the Biological Sciences,* San Diego, Harcourt Brace JOvanovich, 1986.

Clark, J.M.: *Experimental Biochemistry,* New York, W.H. Freeman and Company, 1977.

Dixon, Dougal: *After Man-A Zoology of the Future,* New York, St. Martin's Press, 1981.

Fransman M, Junne G, Roobeek A: *The Biotechnology Revolution?,* Oxford, Blackwell, 1995.

Gaskell G: *Biotechnology-the Making of a Global Controversy,* Cambridge, Cambridge University Press, 2002.

Kurzweil, Ray: *The Age of Spiritual Machines,* New York, Penguin Books, 1999.

Murray, David: *Seeds of Concern: The Genetic Manipulation of Plants,* Sydney, University of New South Wales, 2003.

Old, R.W. : *Principles of Gene Manipulation,* London, Blackwell Scientific Publications, 1989.

Primrose, S.B.: *Principles of Gene Manipulation,* London, Blackwell Scientific Publications, 1989.

Shetty, Kalidas: *Food Biotechnology,* New York, Dekker/CRC Press, 2005.

Smith, John E.: *Biotechnology,* Cambridge, Cambridge University Press, 2004.

Walden, Richard: *Genetic Transformation in Plants,* England, Open University Press, 1988.

Winston, Mark L.: *Travels in the Genetically Modified Zone,* Cambridge, Harvard University, 2004.

Index

H

I

M

N

O

P

R

S

T

V

□□□